THE EARTH

THE EARTH

FROM MYTHS TO KNOWLEDGE

HUBERT KRIVINE

FOREWORD BY TARIQ ALI
AFTERWORD BY JACQUES BOUVERESSE

TRANSLATED BY DAVID FERNBACH

VERSO
London • New York

This English-language edition first published by Verso 2015

First published as *La Terre, des myths au savoir*

1 3 5 7 9 10 8 6 4 2

Verso
UK: 6 Meard Street, London W1F 0EG
US: 20 Jay Street, Suite 1010, Brooklyn, NY 11201
www.versobooks.com

Verso is the imprint of New Left Books

ISBN-13: 978-1-78168-799-4 (HC)
ISBN-13: 978-1-78478-270-2 (Export)
eISBN-13: 978-1-78168-800-7 (US)
eISBN-13: 978-1-78168-798-7 (UK)

British Library Cataloguing in Publication Data
A catalogue record for this book is available from the British Library

Library of Congress Cataloging-in-Publication Data

Krivine, Hubert.
[Terre, des myths au savoir. English]
The earth, from myths to knowledge / Hubert Krivine ; foreword by Tariq Ali ; preface by Jacques Bouveresse ; translated by David Fernbach. – English-language edition.
pages cm
Originally published in French in 2011.
Includes bibliographical references and index.
ISBN 978-1-78168-799-4 (hardback : alk. paper)
1. Geology–History. 2. Earth sciences–History. 3. Religion and science. I. Fernbach, David, translator. II. Title.
QE11.K7513 2015
550–dc23

2014043933

Typeset in Minion Pro by MJ & N Gavan, Truro, Cornwall
Printed in the US by Maple Press

Contents

Foreword by Tariq Ali vii
Introduction xiii
How to Use this Book xix

PART ONE: EARTH'S AGE 1
1 'Pre-Science' 5
2 The Beginning of the Modern Age 13
3 The Twentieth Century and Radioactivity 33

PART TWO: EARTH'S MOVEMENT 47
4 Before Copernicus 51
5 The Construction of Heliocentrism 79
6 Distances 105
7 The Battle over Heliocentrism 113

PART THREE: 'ONLY' SCIENTIFIC TRUTH? 147
8 Why Truth Matters 149

Afterword by Jacques Bouveresse 159
Appendix A. The Proofs of the Earth's Motion 171
Appendix B. Kelvin's Model and Calculation 179
Appendix C. Radioactivity 183
Appendix D. The Copernicus/Tycho Brahe Equivalence 191
Appendix E. The Relativity of Trajectories 195

Acknowledgements 197
Notes 201
Glossary 253
Bibliography 259
Index 271

Foreword

My first thought on reading Hubert Krivine's book was that it should be immediately translated into Arabic, Persian, Urdu, Bengali, Behasa, so that it is available to the new generation that is growing up in difficult circumstances throughout the world of Islam. And not only in those regions. It will benefit many European citizens of Muslim origin, who will find in these pages an open, calm and non-dogmatic interpretation of the origins of this planet and related matters.

Naturally, Krivine challenges all religious orthodoxies that became an obstacle to free thought and especially to scientific knowledge that disputes the ideological foundations of the different religions. The revival of creationism in parts of the US and the bogus concept of 'intelligent design' favored by the unintelligent scribes and politicians (Tony Blair a faded example) are the book's main targets. The growth of religiosity in the West necessitates a response, but this book is even more essential in the Muslim world, where the teaching of biology and natural sciences in state schools is either non-existent or, at best, fragmentary.

That's why I read *The Earth* as an important intervention that could greatly benefit young Muslims everywhere and help transcend both the bombs and the drones from the sky as well as the obscurantist responses from below. In fact, Krivine's text is such a powerful antidote to ignorance and stupidity that it deserves to be a textbook in both the Muslim world and the United States. Forgive the utopian digression. Those who determine what is studied in the educational institutions of most Muslim countries today are either scared by the fundamentalist groups who skulk around in the background or have themselves moved in that direction. Who will educate the educators of the One-Book school? The more open-minded among them will find much of value in this study. It's always better to know what it is that one is disagreeing with.

The presence of an obscurantist layer is not limited to the Islamic world proper. A more sophisticated version has established itself in

Europe and North America, with a few strong voices in the US academy and British political parties. Here the argument sometimes used is that science is a Western imperialist construct and has to be countered with 'Islamic science'. This form of relativism does a disservice to all students, regardless of their faith. And this mode of thought compares unfavorably to the open-mindedness of the scholars who inhabited the Bait ul Hikma [The House of Wisdom] in Baghdad for over four centuries (eighth to eleventh centuries CE), who welcomed debates with scholars from different parts of the world and from different religions, and who were proud to learn from the ancient Greeks and both synthesize and correct that learning with the advances then being made in Baghdad, Cordoba, Palermo and later in Samarqand.

The principle was the same as today: the collection and measurement of data through observation and experimentation. Observatories were constructed with which to study the heavens. The Ptolemaic model was discarded, on the basis of empirical observation rather than religious tradition, by the tenth-century Mesopotamian scientist Ibn al-Haytham, who preceded Descartes and Bacon in insisting that Plato, Euclid and others were mistaken in assuming that the eye produced its own light; the astronomers Nasir al-Din al-Tusi and Ali Qushji seriously debated the possibility that the earth rotated on its own axis. The birth of chemistry, algebra, and the development of geometry was a product of the Arab renaissance. Islamic civilization was the most advanced during that period, and it's hardly a surprise that its confident and assertive scientific intelligentsia produce the finest minds of the Mediterranean world.

Influential Muslim reformers in the imperialism-dominated nineteenth century recalled this period with pride and strongly defended the idea of science separated from both religion and politics. One of them, a militant anti-imperialist, Jamaluddin Afghani, insisted that 'science is continually changing capitals. Sometimes it has moved from East to West and other times from West to East.' He pointed out the dangers of obscurantism and noted how the 'Muslim religion has tried to stifle science and stop its progress. It has thus succeeded in halting the philosophical or intellectual movement and in turning minds from the search for scientific truth.' For him science was universal and he strongly rebuked attempts to force it into religious straitjackets and wrote that

> the strangest thing of all is that our ulema these days have divided science into two parts. One they call Muslim science and one European science ... They have not understood that science is that noble thing that has no connection with any nation, and is not distinguished by anything but itself.[1]

In 1910, Ahmed Kasravi, a young mullah in Tabriz in Persia, stood on the roof of his house and looked upwards. Halley's Comet, the star with a tail, was flaming through the Persian sky. This made him think about the universe. A process of self-questioning began. Kasravi left the seminary and became a historian of his country and a free thinker. In 1946 he was accused of 'slandering Islam', but before he could be tried, a religious fanatic shot him dead. The point of the story is that skepticism can reach the inner sanctums of any religion, as Spinoza and Giordano Bruno proved many centuries ago.

Compare all this to some of the nonsense spouted in the last half of the twentieth century and, of course, today, and the regression is only too visible. During the Zia-ul-Haq military dictatorship in Pakistan (1977–89), when religion was imposed on the hitherto relaxed culture of the country from above, I can recall the transcript of a scientific conference in which a participant was faced with a dilemma. According to Islamic tradition, the Prophet climbed on his horse in Jerusalem (the site of the Dome on the Rock) and flew to heaven for a summit with the Creator. The scientist explained that the Prophet did fly away, but not on a horse: He was transported upwards by a laser beam, a fact that demonstrated the advances of Islamic science. Later, Mr A. A. Abassi, a Pakistani 'neuropsychiatrist', authored a book dangerously entitled *The Quran and Mental Hygiene*, sadly out of print, in which he explained that everything needed in the modern world was already in the Quran, including medicines that cured diabetes, tuberculosis, stomach ulcers, rheumatism, arthritis, asthma and paralysis. Curiously, he excluded mental illnesses. Not to be outdone on this front, a Pakistani nuclear engineer once advised the government that the *jinn* (demons) that appear in the Quran as well as the *Thousand and One Nights* were made by Allah from fire and, as such, could be used as a source of energy to combat the permanent energy crisis that bedevils the country. One can laugh or cry – I prefer to laugh. To explain such eccentricities as a manifestation of hostility to imperialism is simply absurd. Even if it were the case, it could only be the anti-imperialism of fools. The Prophet of Islam is quoted as saying that Believers should seek knowledge wherever it exists, even in far off lands like China. And, in fact, that is where some Pakistani scientists went to study nuclear physics. Others were sent to Europe and the United States. Without this collective knowledge they could not have succeeded in producing the Bomb. The point being is that the idiots in the field should not be confused with the learned.

The father of science in Pakistan, Professor Abdus Salam, shared the Nobel Prize for Physics. Ironically, the Ahmediyya sect to which he belonged was, in 1976, declared non-Muslim by a supposedly secular government led by Zulfiqar Ali Bhutto. It was opportunism at its crudest, and it damaged the country's political culture. Therefore, if one is asked today, 'Has a Muslim ever won the Nobel Prize for any science?' one can answer 'yes' in India or in Britain, but 'no' in Saudi Arabia and Pakistan. There is a further irony in that Salam was a devout Muslim who prayed five times a day but did not allow his beliefs to tamper with his science. There is Divine Truth and there is Reason, proclaimed the twelfth-century Cordoban philosopher Ibn Rushd. Angry clerics burnt his texts at the time, but today it is the philosopher who is remembered.

In Chapter 7 of his book, Krivine explains this duality:

> The crucial question of knowing whether Earth *really* is immobile and the centre of the universe lay at the root of the battle between the Catholic church and Galileo. Like many historical battles, it has often been simplified into a stereotyped image: a backward and uneducated mass opposed to a misunderstood and solitary genius embodying progress.

That is a simplistic view, since to a large extent the struggle was waged within the Church itself, according to its universally accepted rules of the game: the Bible as divine word, and therefore unchallengeable. Until the eighteenth century, scientists (intellectuals) were all believers, even men of the church, if with differing responsibilities. Besides, Galileo was far from being isolated or misunderstood. Just like Kepler, with whom he corresponded, he was honored as one of the leading mathematicians of his time. He was for a long while the protégé of various prelates, in particular Maffeo Barberini, the future Pope Urban VIII, who none the less had him condemned in 1633.

The critique of this naïve view, however, should not lead us to forget that this marked a historical turning point in the manner of conceiving knowledge. Galileo explains in a famous passage in *The Assayer* (1623):

> Philosophy is written in this grand book, the universe, which stands continually open to our gaze. But the book cannot be understood unless one first learns to comprehend the language and read the letters in which it is composed. It is written in the language of mathematics

We are living in bad times again, and good books are as important as ever. It would be a mistake to think that the present is permanent.

Tariq Ali
October 2014

Introduction

> Bernard of Chartres used to say that we are like dwarfs on the shoulders of giants, so that we can see more than they, and things at a greater distance, not by virtue of any sharpness of sight on our part, or any physical distinction, but because we are carried high and raised by their giant size.
>
> – Traditionally attributed to Isaac Newton, but actually from John of Salisbury, *Metalogicon* (1159)

Are we entitled to say that Earth's age is 4.55 billion years, and its trajectory an ellipse centred on the Sun, with an average radius of 150 million kilometres? The majority of educated people today will say yes. Curiously, however, the fact that these assertions constitute what it is customary to call 'scientific truths' is often perceived, three hundred years after the century of Enlightenment, as naïve or even improper. And it is actually *very* educated individuals who say this.

It is beyond the purpose of the present book to trace all the reasons that have led to this pass.[1] We shall just point out one of these, a rational one that it may be possible to influence: choices with a strong political charge have often been presented as resulting from scientific truths that are 'beyond debate'.[2] It is then an easy step to say that it is possible to say anything in science, as a function of one's interests. As for the general public, they know science only through its applications, the worst as well as the best, which is why the euphoria that it generated in the nineteenth century has given way today to scepticism, at least in the rich countries.

This disappointment, reinforced by the observation that scientific progress does not necessarily coincide with social progress, explains the success of a sophisticated relativism, though this has no implication for the actual work of scientific research.[3] The particular status of scientific knowledge is challenged: the scientific approach, like any human construction, does not escape its social determination. There is White science, Black science, women's science, oppressed minorities' science,[4] and so on. But 'science' in the abstract is supposedly a mystification. What needs examination in this context is the revival of various religious fundamentalisms, whether in a caricature form such as creationism, or in the more presentable variant of 'intelligent design'. For Galileo, the book of nature was written in the language of mathematics. For literalists (who hold to a literal textual reading of their holy book), it is set down in the writing of the Bible or the Koran.

What we have here is a renewed obscurantism, expressed with a different emphasis in different contexts. Schematically, in the developed countries (though not exclusively)[5] there is a growing rejection of science; in the poor countries (though again, not exclusively)[6] a rise in religious fundamentalisms.

In the rich countries, the rejection of science is fuelled by the belief that its industrial and military applications are an ineluctable consequence of scientific development. Many of these applications are detrimental to health, employment[7] or the environment, with effects often deemed negative or dangerous. In this case, opposition is healthy. But it becomes sterile if protestors allow themselves to be misled by the propaganda of pressure groups, which justifies their political or social options by supposedly 'scientific' necessities. To accept the responsibility of 'science' in this type of decision thus means renouncing its use in challenging these options. It means in the end capitulating in mid-play by abandoning the claim of rationality to the opposite camp.

In the poor parts of the world, the brutal and constant stranglehold of the leading economic powers arouses a natural reaction of defence on the part of the populations who are their victims. In the lands of Islamic culture, after the retreat of secular nationalist movements, the revival of religious fundamentalism often appears as a radical form of material and cultural resistance. Elsewhere, the proliferation of evangelical sects, despite their different political implications, fulfils the same function: a combination of very real material mutual aid with the demand for dignity or even moral redemption. This is not a regression from the rationality of the century of the Enlightenment, which these countries knew little or not at all.[8] It is rather that this rationality is associated with the supposed 'benefits of Western civilization'.

The present work places itself in a philosophical tradition embodied by Bertrand Russell. It does not lay claim to any original contribution in the field of rationality and the sciences, but – what is perhaps newer – seeks to present conjointly a history of ideas on Earth's age and its movement. The reason for choosing these particular examples is that they cover almost all areas of science (and myth), and thus make possible an analysis of how scientific truths, as distinct from revealed truths, are established and continue to be so. We shall see what price has to be paid, and what precautions taken, to validate a scientific discovery, distinguishing it in this way from an opinion or belief.

We propose therefore, with this object in mind, to trace the genesis of our present understanding of Earth. This is not a treatise on natural science, detailing the battery of techniques needed for this understanding. Nor is it a book of history of science in the strict sense: it largely ignores the socio-historical contexts, focusing 'naïvely' just on the history of ideas, as if their production were simply a rational process opposing pure minds and leading ineluctably from the false to the true without passing through a phase of confusion. In a discussion, after all, it is in principle only ideas that are opposed to ideas.

This book contains two main parts, dealing respectively with the history of Earth's age and the history of its movement. In both of these fields, the development of the right conceptual and technical instruments that made measurement possible is a fascinating adventure. How can such gigantic ages be ultimately measured by ordinary clocks? Or positions and distances with the usual lengths? How can billions of years be related to the scale of a minute, or billions of kilometres to that of a centimetre? The third part sums up the defence of a truth that is 'only' scientific.

Earth's Age

How could Earth have aged by close to 5 billion years in less than 400 years, from the biblical age – which Newton, for example, carefully established as 3998 BCE – to its present age? Scientists such as Kepler, Buffon, Halley, Fourier, Kelvin, Darwin and Rutherford all made their contributions to this. The natural sciences were first called upon to back up biblical teaching (the Flood), but they soon emancipated themselves from this. The burden of proof was turned around: by the late eighteenth century, it was biblical teaching that had to adapt to geology. In the nineteenth century, physics, the queen of the sciences, imposed its conclusions: first, a maximum of 300 million years, then of 20 million. Darwin was almost alone in opposing this reduction. The discovery of radioactivity finally led physicists and biologists to agree on an age of 4.5 billion years. There were no losers: after much resistance, physicists finally confirmed what biologists had correctly glimpsed.

Earth's Position

The history of Earth's position in the universe, or more precisely of its movement, is better known. Medieval astronomy was based on the model of Claudius Ptolemy (second century CE), which accounted for the movements of the planets and Sun *as seen from Earth*: the complicated trajectories of the planets were reduced to circles, and the Sun maintained its position at the centre of the world. The condemnation of Galileo by the Inquisition, for defending the ideas of Copernicus that were 'contrary to Holy Scripture', is well known; the current position of the Catholic church on this condemnation less so.

As against the generally accepted idea, Copernicus' model did not account for the astronomic observations of his time any better than that of Ptolemy. The superiority of the new model is that it intellectually prepared the next step: Kepler, with whom empirical description crossed the threshold of quantity, then Newton, thanks to whom the theoretical basis of planetary movements was finally understood in the light of a mathematical theory of universal movement and gravitation. The essential part of the present conception was established, and modern science was born.

These two first parts may be read independently, though they are none the less connected. The lengths of days and years are measured by the relative movements of Earth and the Sun. The attempt has even been made to date our planet by studying the movement of the Moon in relation to Earth and to provide dates of historic events with the help of a certain particularity of Earth's movement, the precession of the equinoxes (see p. 23).

Traditionally, and particularly in France, science is not part of culture in the broad sense of the term. Not to know that Picasso painted 'Guernica' is unpardonable, but to believe (and even write) that we owe it to Galileo to have shown that Earth is round is commonplace. And yet, the decentring of Earth in the universe that was justified by Newton's law of universal attraction, the age of Earth as fixed by radioactivity, and the theory of evolution initiated by Darwin were decisive steps in modern human culture, whether the fundamental mechanisms of these are known or not. Scientific studies, moreover, too often remain cut off from their own history, which should at least be mentioned by teachers in these respective disciplines. And what about the pleasure that the elegance of an experiment or a proof can provide? The notion of beauty – subjective, to be sure – seems reserved for the world of music, poetry, painting or literature. Scientific knowledge is certainly only *one* knowledge of the world, but it is not the saddest one. These prejudices are all the more damaging in that the culture they create is crippled and partly sterile. Not to mention that they contribute to keeping the public away from the sciences.[9] The present book, which necessarily aims at overcoming this separation of cultures, would thus like to repair that anomaly.

The establishment of a scientific truth is the result of a ferment that mobilizes all the faculties of the mind. The dialogue between theory and experiment, the internal criticism of theories, and the simultaneous quest for the greatest parsimony and the greatest possible universality, mean that this ferment is not a haphazard progress, in which each step is made independently of the previous one. Scientific progress exists as the result of a cumulative process. If it is not a sufficient condition for social progress, it is at least a necessary one.

Hubert Krivine
Paris, June 2010

How to Use this Book

To facilitate a targeted reading of the book, **summaries of the content are given in bold text at the head of each chapter.**

The main conclusions are highlighted in this way.

Several of the endnotes concern related developments and can be read independently.

Finally, a glossary gives the meaning of the main technical terms used.

Printing and, more recently, the internet have made knowledge, and particularly information, tremendously available on a mass scale. They have also, as an inevitable by-product, involved the repetition *ad nauseam* of errors, myths, even mystifications.[1] I have therefore tried, as far as possible, to remain as close as I could to the sources that I have copiously cited.

Except in the Appendices, this book contains scarcely a single equation. I wanted to make it both comprehensible for a non-scientific public and informative for more specialized readers. Only the Appendices, designed for teachers and students in physics or Earth sciences, contain mathematical developments at a sixth-form or university level, even postgraduate (those explaining Kelvin's argument, for example). They can be used by teachers to show how the mathematics taught at secondary school and university is able to resolve fundamental problems.

Finally, the following three diagrams show the eras in which the main scientists discussed in this book lived.

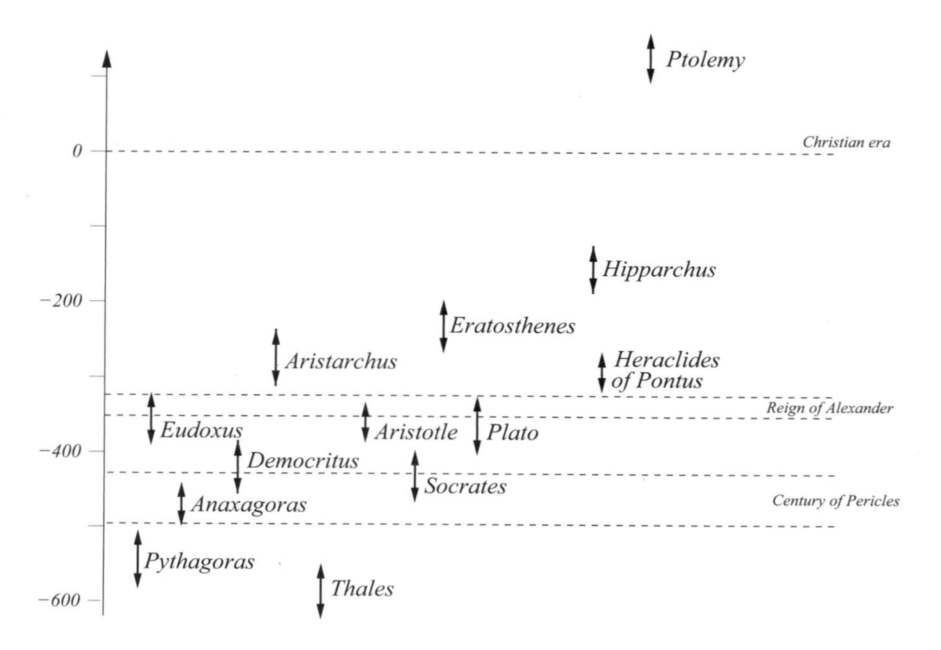

Fig. 1. Greek antiquity

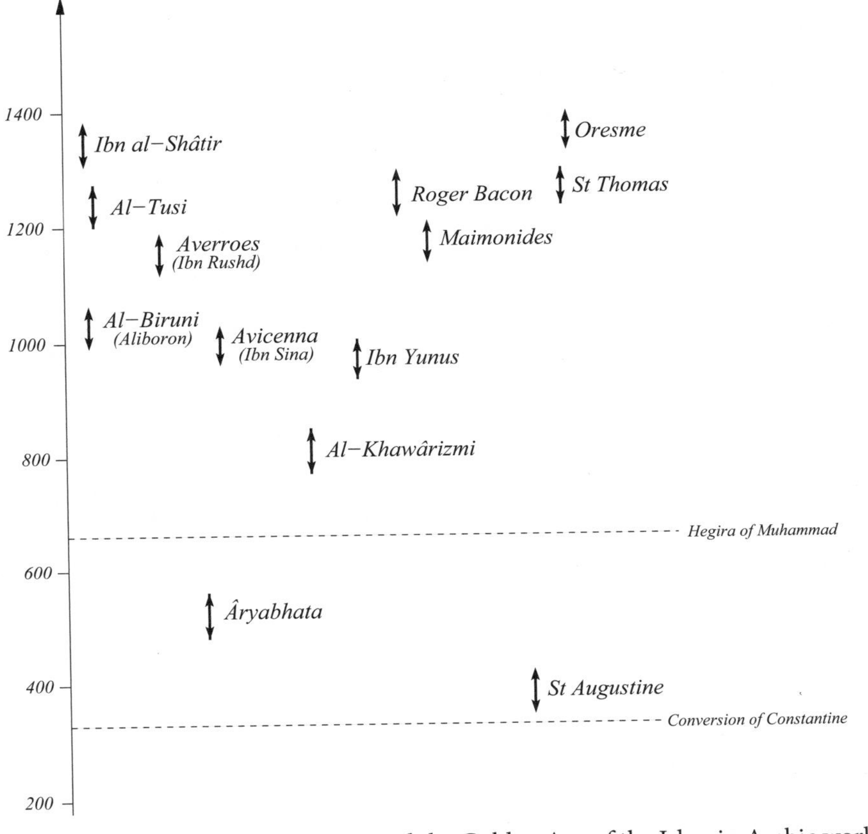

Fig. 2. The European middle ages and the Golden Age of the Islamic-Arabic world

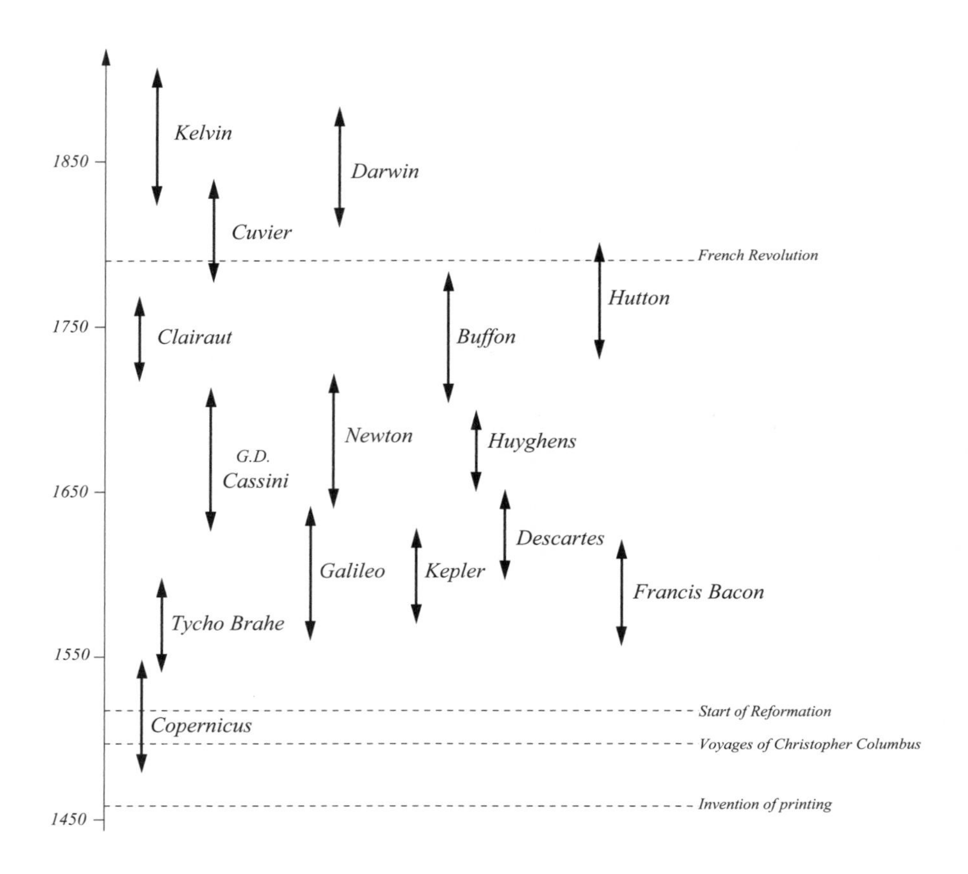

Fig. 3. Renaissance to nineteenth century

Part One

EARTH'S AGE

> The paths by which men come to understand the celestial bodies seem to me as admirable as these bodies themselves.
>
> – Kepler, *Astronomia Nova* (1609)

This part is divided into three chapters, organized not according to chronology but rather to the logic of ideas. The first chapter explains very summarily how the age of our planet was estimated in antiquity, both in Asia and in Europe. It shows how a simple reading of the Bible, considered in Christian Europe up to the Renaissance as the sole source of certain knowledge on this subject, comes up against both internal contradictions (Adam could not have been the first man) and contradictions with known historical facts (the existence of civilizations from before the Flood). In the eighteenth century, the idea took shape that once Earth was created (by God), universal laws of physics could then explain the planet's evolution. These laws undermined the sacred chronology. We shall see[1] how ingeniously different clocks were established that enabled a dating of Earth, as well as the polemics these aroused. Chapter 2 will explain the consequences of the revolution introduced by the discovery of radioactivity, which made it possible to close some major scientific controversies. After a short paragraph on the various scenarios envisaged for the death of the planet, this part concludes with what I see as being the 'moral of this story'.

In the Renaissance, it was generally accepted in Western Europe that Earth's age was round about 5,600 years. Today, within a margin of a few million, it is assessed at 4.55 billion years. How could our planet have gained almost 5 billion years in just a few hundred? The response to this question involves almost every domain of intellectual activity: philosophy, religion, geology, palaeontology, nearly all branches of physics and chemistry (from thermodynamics to nuclear physics), even psychology and, of course, history.

There can be no question here of presenting the mass of knowledge that had to be assimilated to reach this result. We shall rather explain how scientific procedure, the accumulation of hypotheses and results that were refutable, i.e., could potentially be shown to be false,[2] made it possible to reach a truth that is certain enough; in other words, a scientific truth.

A Few Signposts

Before studying the evolution of our ideas on Earth's age, here is a very brief table of the chronology established today.

- The Big Bang happened around 13.7 billion years ago.[3] About this date, viewed as that of the origin of our universe, we shall simply say that for the majority of scientists it marks the moment from which physicists can work (see p. 45). It is hazardous, therefore, to give any meaning to the expression 'before the Big Bang'.
- The solar system and Earth were formed together (within a few million years) some 4.55 billion years ago.
- Bacteria first appeared more than 3 billion years ago: it is now established that certain ancient structures known as stromatolites, currently dated at 3.4 billion years, are of bacterial origin.[4]
- The dinosaurs suddenly disappeared 65 million years ago.
- The appearance of hominids is currently dated at 7 million years: Toumaï supplanted the famous 3-million-year-old Lucy. Still older hominids will undoubtedly be discovered.
- Modern (Cro-Magnon) humans appeared in Europe about 35,000 years ago. We know that they coexisted with Neanderthals. The reasons for the demise of these are still hotly debated.

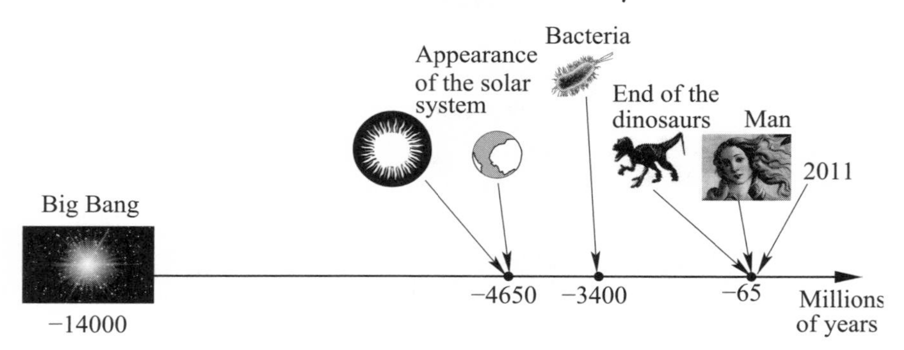

Fig. 4. Timescale from the Big Bang to today (in millions of years)

As shown in Figure 4, the dates for the origin of the Sun and the formation of Earth are identical: likewise, on this scale, the disappearance of the dinosaurs and the year 2015 are indistinguishable.

These results may fluctuate a little,[5] not simply as a function of future discoveries, but also of the definition of the time t = 0. From what moment is it possible to speak of a living organism or even a hominid? How should Earth's origin be defined?[6]

Two conclusions impose themselves. Life appeared very rapidly on Earth, and the lifespan of the human species, *the only one with a genuine power of conception*, is so far derisory on the geological timescale.[7]

We should not forget that for the major monotheistic religions that we shall discuss, Creation meant the (almost) simultaneous creation of the two or three thousand stars visible to the naked eye, the Sun, the planets, Earth, life and man. As we have seen, modern scientists distinguish several timescales, as shown in Figure 4. But what about the zero point on this scale? There is a naïve – and very widespread – view of the date of the Big Bang as that of the appearance of the universe. Instead of the biblical dating of a rather skimpy universe, the Big Bang gave rise, as it were, to today's universe made up of thousands of billions of galaxies. Let us simply say that we know now that the universe is expanding. And so, if we go far enough back in time, we reach a situation in which it was extraordinarily dense and hot. The theory of relativity makes it possible to look back and set dates. We can add that many observational results confirm the existence of this initial cauldron, showing traces of it today. The problem is that the same theory of relativity tells us that in a milieu of such density and energy, the notion of time completely loses its intuitive character. Besides, it is not just the theory of gravity that we have to apply, but also quantum mechanics, and we still do not know how to operate these simultaneously. For this reason, the majority of scientists refuse to answer the question of the date of the universe; it is not even certain that we can define the meaning of this question.[8]

CHAPTER 1

'Pre-Science'

Many 'primitive' societies have no notion of Earth's origin. For some civilizations it has always existed, unchangeable; for others, it has undergone a cyclical development over a very long period. The Bible introduces not simply the creation of the world, but also a very precise chronology.

The myths according to which Earth (Gaïa) was born of Chaos seem to have had only little influence on the development of Greek philosophy.[1] In a captivating essay, Paul Veyne makes the point that religion (and philosophy?) maintained 'an extremely loose connection' with mythology: Aristotle certainly believed in the gods, but to a large extent his physics dispensed with them.[2]

The Greek atomists accepted an origin for Earth, but not starting from nothing. For Epicurus (third century BCE), heir to the materialism of Democritus (460–370 BCE), 'in the beginning' atoms all fell in parallel, with a uniform motion. Then one of them underwent a small deviation[3] and thereby collided with the others. This is the *clinamen* that Lucretius (first century BCE) describes[4] in his famous *De rerum natura*:

> Atoms indeed descend in a straight line in the void, drawn by their weight; but it happens, one cannot say where or when, that they swerve somewhat from the vertical, so little that it is scarcely possible to speak of declination. Without this swerving, none of them would cease falling through the immense void, like drops of rain; there would be no meetings or collisions of any kind, and nature would never have been able to create anything.

From this point on, triggered by these 'interwoven atoms', a snowball effect starts to create the various bodies.

Aristotle[5] (384–322 BCE), who opposed atomism, was not only a major figure in antique philosophy; following his rediscovery by the

Arabs in the eleventh century,[6] he influenced Christian culture in the Middle Ages.[7] Aristotle distinguished a supralunary world (located beyond the Moon), unaltered and unalterable,[8] from a sublunary world that was changing and corruptible. In his treatise *On the Heavens*, book 2, part 1, he wrote:

> After everything that has been said, therefore, we can see clearly that the heavens as a whole were not created, and no more can they perish, as some philosophers say, but that they are one and eternal, having neither a beginning nor an end, for all eternity.

For Aristotle, Earth, though subject to constant changes, remained fundamentally immutable. It had always existed.

In contrast to this view, the religions based on the Bible, and later the Koran, marked a radical break in their assertion of a genesis for Earth. This break was not absolute: their accounts show a striking affinity with texts found on Babylonian tablets.[9] These tablets, which date from the thirteenth century BCE, tell of a separation between Heaven and Earth, the appearance of man, and later a flood of seven days and seven nights, brought about by the gods to punish humanity. Only Utanapishti escaped: he was warned and thus able to build an ark for himself, his family and his animals. Like Noah, the end of these trials was signalled by a dove that brought him an olive branch. We also find here the story of the goddess Ninti (Eve?), drawn from a rib of Enki (Adam?), who was punished for eating the plants in the garden of Dilmun (Eden?).

In the Middle Ages, the age of Earth was not a fundamental question: Jean Buridan, a major thinker of the fourteenth century – even if he has not always enjoyed a favourable opinion – wrote:

> I also conjecture that the world has existed in perpetuity, as Aristotle appears to have understood it, … even if it is false according to our faith.[10]

This flexibility in relation to the sacred texts may seem remarkable to us today. But this is a result of the major change in mentality imposed by the Counter-Reformation two centuries later.

The great Muslim philosopher Averroes (1126–98), brought up on the thinking of Aristotle, likewise rejected creation *ex nihilo*. But wishing all the same to remain faithful to the Holy Book, he cleverly sidestepped the problem:

> For in actual fact, the divine utterance: 'It is He who created the heavens and Earth in six days – His throne was then on the water',[11] maintains, by its self-evident meaning, that there was something that preceded this existence [i.e., that of 'the throne' and 'the water'], and that time passed previous to that time.[12]

The majority of Islamic philosophers of this time sensed the same tension between sacred text and philosophy.[13] Maimonides, for his part, inclined against the eternity of the world. For him, the question was not decidable by reason, so it was better to stick to tradition.

The upheavals of the Renaissance affected all fields of knowledge, often in interacting ways. The 'discovery' of America in 1492, and the great voyages of the Portuguese navigators, transformed the image of the world; developments in firearms and printing (initially conceived solely to facilitate the reproduction of the Bible)[14] ensured European supremacy in Africa and the Americas. Francis Bacon[15] (1561–1626) expressed this transformation quite clearly:

> Printing [*artis nimirum Imprimendi*], gunpowder [*Pulveris Tormentarii*] and the compass [*Acus Nauticae*]. These three inventions have changed the face and the state of the world. The first in literature, the second in the art of war, and the third in navigation: they have produced countless changes of such a kind that no empire, sect or star has exercised more power and influence in human affairs than these mechanical inventions.[16]

In the sphere of culture, scientific reflection, represented in particular by the heliocentrism of Copernicus (1473–1543), then by the astronomy and physics of Galileo (1564–1642), was accompanied by an artistic renovation. At that time, in contrast with today, scientific knowledge was part of general culture. The initiatives of Copernicus or Kepler can be understood only in the light of the general knowledge and beliefs of their day.[17]

As far as the age of Earth is concerned, correction would only come later: there was no reason to doubt that, as the Bible taught, the universe, Earth, life and man – and woman – appeared almost simultaneously. As the Holy Book was also treated as a book of history, it was possible to determine absolute dates. It was to this 'monastic' work that the greatest minds of the time applied themselves. They took the generations listed in the Bible (starting with Adam, Eve, Cain, Abel …), added a few astronomic calculations about eclipses and the precession of the equinoxes (see below, p. 45) and arrived at the following dates for the origin of the world: 3993 BCE, according to Kepler (1596); 3998 BCE, according to Newton;[18] 4004 BCE, according to the Anglican archbishop James Ussher, and even the 23rd of October![19]

These were only the most famous authors, who never doubted their calculations. And yet, as we know, this did not prevent Kepler and Newton from revolutionizing astronomy in the seventeenth century. Bishop Ussher, for his part, was viewed for four hundred years as the pope of biblical dating, even of dating in general. The concordance[20] between these calculations, each carried out independently, was sufficiently convincing. For the two later writers in particular, this involved a fantastic work of erudition. In actual fact, the Bible only gives a detailed chronology of the earliest genealogies: Adam lived for 930 years, having a son Seth at the age of 130 years, who produced Enosh at 105 years, who had Kenan at 90 years, etc. This then makes it easy to calculate the date of Noah's birth: 1,056 years after the Creation. As Noah was 600 years old when the Flood came, this would then be dated 1,656 years after the Creation. Abraham was born 292 years later. Up to this point, there is total precision,[21] but from then on the chronology gets far more vague. Recourse is then needed to pagan history, supposedly truthful, from the reign of Nebuchadnezzar II (second book of Kings) in the sixth century BCE.

In fact, hundreds of calculations had already been undertaken on this basis; Fuller offers a histogram showing 156 dates varying from 6500 BCE to 3600 BCE.[22] The histogram presents two marked peaks, around 5500 BCE and 4000 BCE respectively, depending on the version of the Bible (Vulgate or Septuagint) used by the calculators.

Before commencing on the 'scientific' critiques of the Bible, we should mention that factual contradictions had also been noted.[23]

Isaac de la Peyrère, a protégé of Queen Christina of Sweden (who also hosted Descartes), published in 1655 a rather muddled text, *The Pre-Adamites*,[24] in which he argued that Adam was not the first man: how did Cain manage to find a wife, and how could God tell men to kill him if the only people were he and his parents? The conclusion was that Adam was simply the first Jew, the sole bearer of original sin, and that the world was far older than presented in the Bible. La Peyrère's constant – and generous – idea was to facilitate the conversion of the Jews to the 'true' religion.[25] In the end he had to retract under pressure from Pope Alexander VII, but he still continued his research.

In the same era, the Jesuit missionary Martino Martini (1614–61) investigated the question. The chronology of Chinese dynasties that he brought back seemed firmly established and, unlike the Bible, without any gap; it went back six hundred years *before* the Flood, which it did not mention.[26] So the Flood could not have been universal. The pope retorted that the Chinese, out of vanity, had fiddled with the dates in order to make their people the most ancient. The edition of Martini's *History of China* published thirty years after his death was preceded by a warning:

> He [Martini] rejected the vanity of the Chinese opinion of the antiquity of the world and did not deem it any more prejudicial to the Christian chronology than the idea of the Indians who claimed to be far older than the Moon.[27]

Martini's own text, however, was more ambiguous:

> And if we credit their historians, we would necessarily have to believe that the origin of the world preceded the Flood by several thousand years, as has already been remarked. It is true that they hold the continuation of the chronology to be beyond debate, from the reign of the first emperor on, and that there is no nation on earth that is so diligent as the Chinese, or so well instructed in the knowledge of time … Each person will freely take the view here that seems more likely to him; but the conclusion is based on creditable proofs.

There follows a very precise enumeration: Fohius (Fou Hi) began his reign in 2952 BCE, Xin Ung in 2837, Hoangti in 2697, Xaoha in 2587 and so on.

And indeed, eighty years later the Chinese Jesuits verified the dates of the eclipses carefully listed in the *Spring and Autumn Annals*, one of the Chinese 'five classics';[28] these coincided exactly with the results of astronomic calculation. That seemed to confirm the Chinese dating.

We know today that Chinese chronology had no better foundation than that of the Bible or the tales of the *Iliad*;[29] from this point of view, the pope was not wrong. But the Jesuits believed it all the same, and even went on to propose using the Septuagint version of the Bible rather than the Vulgate, with a view to adjusting the date of the Flood and gaining an additional ten centuries.[30] At all events, doubt had been cast on the correctness of the Mosaic chronology. This aroused a violent polemic in the second half of the seventeenth century. Completely rational arguments were advanced, such as the following: the Vulgate chronology leaves only a hundred years between the end of the Flood and the construction of the tower of Babel, and since Noah was the only survivor, his descendants would not have had time to multiply sufficiently to build this structure; moreover, 'the Jews corrupted the Hebrew text' (whose chronology was taken over by the Vulgate) in order to deny Christ his quality as Messiah by putting back the prophesied[31] date of his appearance.[32]

There was an active rivalry between the Anglican or reformed churches, rigorous upholders of the Hebrew Bible, and the Catholics, who were more open. At all events, the Jesuits battled for the adoption of the Septuagint bible, compatible with Chinese chronology, arguing that this would facilitate the conversion of this immense people to the 'true' religion and strengthen the connection between the Celestial Empire and Louis XIV.

The stimulating argument of Virgile Pinot, a great authority on this matter, contains a detailed account of the influence of the Catholic missions in China. But she also emphasizes the reciprocal effect :

> As a consequence of this movement [the discussions as to the authenticity of Chinese chronology] before 1740, people [in the Christian scientific milieu] were led more to doubt the universal Flood than the antiquity of the Chinese chronology.[33]

On the historical account itself, Finkelstein and Silberman offer in *The Bible Unearthed* a list of internal contradictions that were already denounced[34] in the seventeenth century.[35] Despite being academic, this book has played a political role in Israel comparable with that of the new historians who challenge the official historical view of the birth of the state of Israel.[36]

Finally, Champollion's[37] deciphering of the detailed history of the Egyptian dynasties, which does not mention the Flood, raised the same problems for the pope in the 1830s.[38] In fact, despite being known as the 'Robespierre of Grenoble', Champollion had rather started out with what he himself called an odour of sanctity in his relationship with Pope Leo XII, to the point that the latter had proposed a cardinal's hat for the 'saver of biblical chronology'. Cuvier likewise expressed his congratulations.[39] Champollion had in fact resolved the celebrated dispute over the Dendera zodiac, by dating this bas-relief to the Greco-Roman era, contrary to the opinion of his colleagues who believed it antediluvian. Little concerned by this distinction,[40] Champollion confessed in a letter to a friend:

> Here I am, viewed by the devout party as a Father of the Church. I am tired of this odour of sanctity that the devout are starting to find with me, as if I had no other goal in working for fifteen years to discover my alphabet than the glory of God ... But I shall see them make quite a nasty grimace one day, when I develop the immediate consequences and conclusions of my discovery. He who laughs last laughs loudest.

We should note that 'Champollion had been asked by Rome to give a written assurance that the future developments of his science would not damage the authority of the Bible'.[41] That was far from the case, and his subsequent dating of the earliest Egyptian dynasties went back well before the Vulgate's date for the Flood.

CHAPTER 2

The Beginning of the Modern Age

The seventeenth century saw the start of a cautious process of taking distance from the Holy Scriptures and applying the laws of physics to the study of Earth's history. The initial idea was to verify the reality of the Flood, as attested to by marine fossils found even in the high mountains. The difficulty was to find objective traces of a past era. How were suitable clocks to be determined?

The idea that the laws of physics are universal, applicable not only to the present but also to past and future, may seem commonplace today, or at least we hope so. But this could not be taken for granted, even in the eighteenth century.[1] Descartes (1596–1650) was one of the initiators of this idea, even a militant pioneer, breaking deliberately with a certain elitist tradition.[2] For him, divine intervention was limited to the Creation; the laws of physics governed the subsequent development of valleys, rivers and mountains. Pascal's *Pensées* give a good summary of this philosophy:

> I cannot forgive Descartes. In all his philosophy he would have been quite willing to dispense with God. But he had to make Him give a fillip to set the world in motion; beyond this, he has no further need of God.

This idea had a devastating effect on biblical chronology. Thomas Burnett (1636–1715), for example, attempted to reconcile biblical time with the time of physics. The six days of Creation were indeed rather short, but was it not written in the second letter of St Peter:[3] 'with the Lord one day is like a thousand years and a thousand years like one day' (2 Peter 3:8)?

In a letter of 1680, Newton[4] deployed an astute approach to rescue the biblical dating, in his courteous argument with Burnett – showing a delicacy that was not so customary with him.[5]

> Now for ye number & length of ye six days: by what is said above you may make ye first day as long as you please, & ye second day too if there was no diurnal motion till there was a terraqueous globe, that is till towards ye end of that days work.

A century later, Buffon wrote in the same vein:

> What may we understand by the six days that the Holy Author described for us so precisely, counting them one after the other, if not six spaces of time, six intervals of duration? And these spaces of time indicated by the name of days, for want of another expression, can have no relationship with our present days, since three of these days passed in succession before the Sun was placed in the heavens.[6]

The biblical method had the unchallengeable advantage of providing an absolute date – even if it required the help of a little astronomy and interpretation. What laws of physics would be able to offer the same possibility?

Emancipation from the teachings of Genesis was far from being sudden. The reality of the biblical Flood continued to haunt scientists until the first half of the nineteenth century. But was it indeed healthy to seek to *prove* the Flood? We may appreciate the wise opinion of Johann Gottlob Krüger:[7]

> The story told by Moses speaks of a universal Flood. We are obliged to believe this, and we would do well to keep it at that. But an itch to prove everything leads us to seek arguments to verify a fact that no one doubts; and lacking evident signs of this, we make ourselves demonstrators of sacred history for the common cause of religion. I do not blame the intention; but experience shows that this indiscreet zeal often has very dangerous outcomes, for those who simply believed the fact without demanding proofs would have always continued to believe, and those who were disposed to doubt it will certainly be maintained and even confirmed by seeing the insufficiency of our arguments to convince them.[8]

Buffon, who was far from being a bigot, curiously seems to have taken the sacred text quite seriously:

> Burnet, Whiston and Woodward have committed a fault that we see as deserving to be corrected, which is to have regarded the flood as possible by the action of natural causes, whereas Holy Scripture presents it to us as produced by the immediate will of God; there is no natural cause that could produce on the entire surface of the Earth the quantity of water that would have been needed to cover the highest mountains; and even if one were able to imagine a cause proportionate to this effect, it would still be impossible to find a further cause capable of making the waters disappear.[9]

Buffon strongly emphasizes this material impossibility of the Flood.[10] To the champions of a (bad) physics, he rather remarkably opposes the sacred text. And by declaring it impossible to apply to the Flood the ordinary laws of physics, he breaks with Descartes's programme and his natural philosophy. Unless, if we venture this hypothesis, Buffon was not completely in good faith, and sneakily used the Bible to attack the upholders of a scientifically established Flood, and hence of any such thing at all,[11] as he did not believe it. This was a man who, after all, did not hesitate to draw the wrath of the church with his 'non-Catholic' estimate of Earth's age. But Cuvier could still believe in the biblical Flood in 1830, even if he ended up dating it some 70,000 years before Moses' time.

The First Clocks

The eighteenth century saw the origin of ideas that would (almost)[12] all be further developed:

1. Times of stratification and erosion.
2. Cooling times for spheres of different sizes, extrapolated to Earth.
3. The salt content of the oceans.
4. The changing distance of the Moon from Earth.
5. Finally, though only for historical time, the precession of the equinoxes.[13]

1. Times of stratification and erosion

From the second half of the seventeenth century, the discipline of stratigraphy, i.e., the study of the successive deposits of material, underwent a notable development. But if this made it possible to conclude (almost) certainly the simultaneity or anteriority of events, it remained far vaguer in yielding figures, let alone offering absolute dates. For this reason, scientific societies for a long time rejected communications claiming to date Earth,[14] as they did those dealing with the origin of languages: these problems seemed outside the grasp of rational knowledge. Stratigraphic dating rested on hazardous assumptions of calculation: if it took a hundred years to deposit a millimetre of clay, and if a stratum measured one metre, then the time needed for the deposit was 100,000 years. Times of erosion, for the carving of valleys and canyons, were studied with the same method. The Danish scholar and bishop Niels Stensen (1638–86) (also known as Nicolas Steno), who was convinced of the biological nature of fossils, can be considered one of the founding fathers of stratigraphy. We should note, however, that he did not draw from this any conclusion as to time scales, and remained faithful to biblical teaching on this question.

The origin of fossils was long debated: were these a trick of nature, like crystals or desert roses, or were they the traces of extinct animals? Ideas on this question evolved in a complex way; at the end of this book we reproduce an extract from the *Encyclopédie* that gives an excellent summary of the question.[15]

Benoît de Maillet (1656–1738) was a precursor of Buffon, rather as Giordano Bruno was to Galileo. He was an eccentric and a great admirer of Cyrano de Bergerac, to whom he dedicated an extremely well-documented book. His key contention was that all land had arisen from the sea. He extrapolated the speed of retreat of the sea, particularly on the basis of ancient ports that were now on dry land, and arrived at an age of 2 billion years.[16] He also used the height of mountains:

> If it is true that the sea decreases, it is no less so that it is by no means impossible to find the correct measure of its present decrease. Now, comparing this decrease with the height of our highest mountains, can we not likewise take the measure of the time that the sea has taken to decrease from this height to its present surface, and accordingly know the number of centuries since our globe became habitable?

This extract may lead one to ask whether de Maillet, like all his contemporaries, did not identify Earth's age with that of humanity. Fearing the wrath of the church, he took the triple precaution of writing under a pseudonym (Telliamet, a near-reversal of de Maillet), not being published until ten years after his death and, what is more, only then in the Netherlands. He gave a convincing demonstration of the biological origin of fossils. For him, coal arose from decomposed and compressed trees, and volcanoes burned with the fat of whales. With the same enthusiasm, he gathered evidence of the present existence of fish-men (and -women), as a basis for his hypothesis that everything comes from the sea. He also has a very original way of developing the question of the Creation:

> In actual fact, to draw on the thinking of one of your authors ([Montesquieu] *Persian Letters*, letter 109), can those who know nature and have a rational idea of God understand that matter and things were created only 6,000 years ago; that God delayed his work for the whole of the preceding eternity; and that he only used his creative power yesterday? Was this because he could not have done so, or because he did not want to? But if he could not have done so at one time, he could not have done so at another. So it is because he did not want to. But as there is no succession in God, then, if we accept that he wanted one thing on one occasion, he wanted it always, that is, from all his eternity.[17]

However fantastical this may seem in certain aspects, it was drawn on by both Buffon and Cuvier.

Buffon (1707–88) was also interested in times of sedimentation. Impressed by the thickness of the sedimentary strata in the Alps and the slowness of deposits formed by the oceans, he ended up with an age of several million years, or even 3 billion.[18] He was prudent enough not to publish these results.

In the nineteenth century, thanks in particular to Cuvier (1769–1832) and William Smith (1769–1839), who had the idea of dating geological strata by the fossils they contained, stratigraphy became the main source of information on Earth's history. It made it possible to compare ages, but arriving at an absolute age[19] still depended on estimating erosion times. Darwin was led to go back on his own famous calculation of the age of the hills of the Weald, in Kent, which he estimated by this method at 300 million years, thereby providing a lower limit to Earth's age.

2. Cooling times

Buffon's calorimetric experiments

Buffon can be seen as the father of scientific dating. What he proposed, in fact, was both theoretical models and corresponding experimental measurements for these. His model of sedimentation has already been mentioned, but he also used a second method, assuming that today's Earth is the result of the cooling of a planet initially composed of molten rock. His hypothesis was that of an incandescent sphere (which defined time $t = 0$) that steadily cooled. This led him to conduct an experiment. In the Montbard ironworks he raised to a red-hot state spheres of different radiuses and made of various metals, then measured the times they took to cool to ambient temperature. These results he extrapolated to a sphere of terrestrial dimensions. But he had no theory for doing this, and his extrapolation from balls of between 0.5 and 5 inches to the terrestrial radius of 6,400 km was false. It gave him an age[20] in the order of 74,000 years.[21]

However ridiculously brief this may seem to us, it went far beyond that deduced from the Bible. Summoned in 1751 by the deputies and syndics of the theology faculty to retract his 'propositions contrary to the belief of the Church', Buffon complied without hesitation. He later wrote to the president, de Brosses: 'I feel that I shall be obliged to suppress the few good things that there were to be said; but in the end, as someone said, it is better to be flat than hanged.'[22]

We are far here from the '*Eppur si muove*' ('and yet it turns') attributed to Galileo.

The theory of Kelvin (1824–1907)

Buffon's theory was subsequently 'rescued' by William Thomson, later Lord Kelvin. He was certainly one of the greatest physicists of his time, and enjoyed immense prestige. Basing himself on the equation for heat,[23] he initially gave a spread of 20 to 400 million years for Earth's age, in 1862, but reduced this to 20 to 40 million in 1897.

Given the very major – even dominant – impact that this calculation had throughout the second half of the nineteenth century, it is worth dwelling on it a while. Appendix B below explains Kelvin's mathematical work in detail. He started from the same model as Buffon: what he called the 'beginning of the Earth' (initial conditions) was a sphere of molten rock at a uniform temperature estimated at 3,900°C. The temperature of its surface, in contact with the vacuum outside (what are called limit conditions) stabilized at a reasonable level of around 20°C. We know this because life began developing a very long time ago, and required a stable temperature of this level. At Earth's centre, on the other hand, the temperature remained fairly close to its initial value due to thermal inertia. Between Earth's surface and its interior, therefore, there was a temperature continuum rising from 20°C to 3,900°C. We know today that in probing beneath Earth's surface the temperature rises by around 3°C every 100 metres (viz. the temperatures in deep mines).[24] This is what is known as the *temperature gradient* at Earth's surface, which changes with depth. At Earth's origin the gradient was far steeper, almost infinite: very rapidly, that is, over a very short distance, there was a transition from the (low) surface temperature to the (high) temperature of the core; then the cold slowly reached farther down and the gradient declined, stabilizing at its present value of 3°C per 100 metres. It was shown that the way in which this gradient declined over time could be determined theoretically, thanks to Fourier's equation: if the initial conditions and the limit conditions were known, it was possible to deduce the time needed to lower the temperature gradient to its present value. The use of Fourier's equation (see Appendix B) required only that a constant κ was known.[25]

To be complete, we need to add that this equation only supplies the evolution of Earth's temperature if we assume it to be rigid, that is, without any possible internal movement of matter. Kelvin was aware of this,[26] and believed that he had demonstrated the validity of his hypothesis by astronomical considerations:

> It seems therefore nearly certain, with no other evidence than is afforded by the tides, that the tidal effective rigidity of the earth must be greater than that of glass ... Even if the rigidity were as much as that of steel, the precession and mutation would not be more than three-fifths of their full amount for a perfectly rigid spheroid.[27]

Taking into account the uncertainties as to the initial conditions, the limit conditions and κ, Kelvin arrived in 1862 at a spread of 20 to 400 million years. He wrote:

> But we are very ignorant as to the effects of high temperatures in altering the conductivities and specific heats of rocks, and as to their latent heat of fusions. We must, therefore, allow very wide limits in such an estimate as I have

> attempted to make; but I think we may with much probability say that the consolidation cannot have taken place less than 20,000,000 years ago, or we should have more underground heat than we actually have, not more than 400,000,000 years ago, or we should not have so much as the least observed underground increment of temperature. That is to say, I conclude that Leibnitz's epoch of 'emergence' of the '*consistentior status*' was probably between those dates.[28]

We have to understand why this assertion was taken as gospel. The validity of Fourier's equation, which was still being successfully tested, seemed impossible to fault; it had almost the same authority as the law of gravity.

Next, to reduce his range and verify its coherence, Kelvin calculated the age of the Sun, which was necessarily older than our planet.[29] He had first to understand the Sun's mechanism of combustion. It was known (see Appendix C, p. 183) that this could not be a chemical reaction: a Sun burning carbon, for example, would last only a few thousand years. Kelvin attributed the Sun's energy to the only possible source known at this time: gravitational energy, that is, the potential energy lost by the gradual collapse of the Sun due to its own weight and transformed into heat. He obtained an age for the Sun 'probably lower than 100 million years':

> It seems, therefore, on the whole most probable that the sun has not illuminated the earth for 100,000,000 years, and almost certain that he has not done so for 500,000,000 years. As for the future, we may say, with equal certainty, that inhabitants of the earth can not continue to enjoy the light and heat essential to their life for many million years longer unless sources now unknown to us are prepared in the great storehouse of creation.[30]

Since Earth is younger than the Sun, this gives an upper limit of its age. Other studies on the terrestrial crust led to a further restriction of the original range to 20–40 million years, which Richet pleasantly calls Kelvin's garrotte.[31]

We should note Chamberlin's (1843–1928)[32] intuition, writing four years before the role of radioactivity was understood:

> Is present knowledge relative to the behaviour of matter under such extraordinary conditions as obtain in the interior of the sun sufficiently exhaustive to warrant the assertion that no unrecognized sources of heat reside there? What the internal constitution of the atoms may be is yet an open question. It is not

improbable that they are complex organizations and the seats of enormous energies. Certainly, no careful chemist would affirm either that the atoms are really elementary or that there may not be locked up in them energies of the first magnitude.[33]

3. The salt content of the oceans

Halley's idea[34]

Halley explained that the salt content of the sea was introduced by the fresh water of rivers. This is only an apparent paradox. The supposedly fresh water of rivers actually contains various salts (of sodium, potassium, calcium ...) that come from the rocks it has flowed over (you need only read the labels of mineral water bottles!). It continuously supplies these salts to the ocean, which constantly evaporates. But the evaporating water vapour is completely pure, so the outcome is clear: the water left in the oceans is steadily charged with river salts. In support of this thesis, lakes that have an outflow are not salty, but they become so if they are isolated. And so by estimating the quantity of salts in the oceans and the total outflow of rivers (in tonnes of salt per year), it is possible to deduce the time needed to contribute these.

This is a highly simplified model, but still representative of scientific procedure: a theory, and an experimental possibility of validating it. Halley did not offer an estimate, but believed his model would prove that Earth was far older than was credited. More than a century later, this idea was exploited.

The measurements of Joly (1857–1933)

Without initially being aware of it, John Joly rediscovered Halley's idea a century later, and deduced from it a time in the order of 100 million years. He estimated the mass of sodium contained in the oceans at 1.41×10^{16} tonnes, and the rate of annual deposit from rivers at 1.57×10^{8} tonnes; the ratio of the two gives an order of magnitude for the age of ocean formation at 90 million years.[35] That of Earth is naturally greater. This model of sodium contribution by rivers alone and without loss is very simple but mistaken: there are processes of salt loss (winds, deposits), with the result that the salinity of the oceans has in fact varied only very little in the last 1 or 2 billion years. But this estimate still played a role at the start of the twentieth century.

4. The distance from Earth to the Moon and the length of the day

The Moon, which initially came from Earth, is moving away from it.[36] Armed with a model of this changing distance, and going back in time, we can get an idea of the date of separation, and from this an estimate of a minimum age for Earth. The astronomer George Darwin (1845–1912), a son of Charles, tackled this thorny problem.

Darwin junior's calculation

George Darwin undertook an analytical study of this complex problem of the Earth–Moon dynamic, taking into account the Sun and the other planets. He assumed, which we know now to be false,[37] that the rotating Earth, still hot and soft, was distorted by centrifugal force and that the mass that would form the Moon then broke away. Even if we assumed that the Earth–Moon system was isolated (neglecting the influence of the Sun and the other planets), the problem would still be complicated. In actual fact, all parts of Earth and all parts of the Moon are mutually attracted by the force of gravity, with an intensity that depends[38] on their respective distances. The portions of Earth closer to the Moon are more attracted by it than the points at the antipodes. These different intensities, applied to the same body, create tensions even capable of breaking it. They are responsible for the tides. Tides exist on the Moon as well as on Earth, exerting their effect on both solids and liquids, and distorting them. We can understand why the swellings thus formed are greater on liquids. These swellings, which would be located on the axis connecting the centres of the two planets if they were stationary, are slightly shifted: Earth, rotating on its axis in 24 hours as against the Moon which turns around Earth in 28 days, brings about this swelling which, by friction, dissipates energy and so brakes Earth's rotation on itself.[39] A direct consequence is that the days are getting longer. A mechanical consequence[40] is that the Earth–Moon distance is increasing, as we have said. At the time, George Darwin could not have known this protraction by experiment, but he was able to estimate it theoretically.

Darwin's calculation gave 56 million years for the birth of the Moon, an order of magnitude compatible with the estimates of Kelvin, who adopted these results, but in contradiction with those of his father.

The friction caused by the tides slows down the rotation of Earth on itself, so there is a relationship between the length of the day and the cycle of the tides. The attempt has been made to calculate the periodicity of this cycle in the past by studying the alternating strata of sand and silt left by the tides in certain estuaries.[41]

Another track is the study of stromatolites, which can yield direct information. The fact that stromatolites, which have existed for billions of years, are the product of living creatures, and thus far more sensitive to the vagaries of the environment than is mineral matter, suggested to Vanyo and Awramik their use[42] as a clock. The oldest, dated some 3.5 billion years ago, carry traces due to the variations of activity of the cyanobacteria that compose them. Counting these gives the number of days in the year, rather in the way that the rings of trees indicate their years of life (see p. 67, below).

Finally, we should cite an ingenious method of determining lunar cycles by studying the growth of nautilus shells.[43]

5. The precession of the equinoxes

Use of the precession of the equinoxes is valid only for historical time, since it assumes that we know the Sun's reported position in relation to the constellations over the course of the year. It was used by Newton (see chapter 1, note 18) in his attempt to connect sacred history with history in general. The principle is as follows: We know that Earth turns around the Sun (in a year), in an ellipse whose plane is known as the ecliptic; it also turns on itself (in 24 hours), round the axis of the poles. This axis

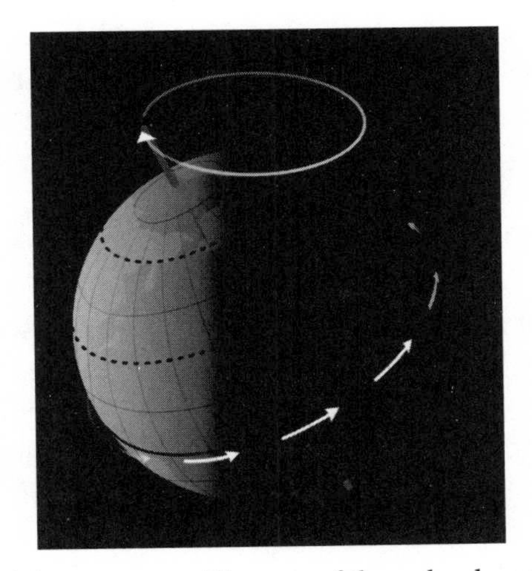

Fig. 5. Precession of the equinoxes. The axis of the poles describes the white circle in 26,000 years.

is not perpendicular to the plane of the ecliptic, but makes an angle of 23° with this, and it itself rotates over a period of some 26,000 years.[44] This means that, if the axis of the poles currently points towards the star Polaris, in 3000 BCE it was rather towards the star Thuban, in the Dragon constellation. For a terrestrial observer, the view of the sky has changed. We thus understand how the precise astronomical measurements of the ancients make it possible to date the moment at which they were taken.

The Great Quarrels

By the nineteenth century, the long timeframes of the geologists were generally accepted. But polemics remained, opposing 'catastrophists' to 'uniformitarians', 'Plutonists' (who proposed the motive role of fire) to 'Neptunists' (who proposed the driving role of water), believers in a limited age to those believing in an immense or even infinite time (Kelvin against Darwin). Often, though not always, catastrophists were to be found in the same camp as Neptunists, with uniformitarians and Plutonists in the opposing camp.

Catastrophists versus uniformitarians

Without going back to the myth of Atlantis, there can be no doubt that the Flood, very present in the Bible and seemingly authenticated by the universal presence of marine fossils, provided a cultural background for both Neptunists and catastrophists.

Catastrophists

For this school, Earth today is the result of cataclysms (comets, volcanic eruptions, Flood or floods, etc.). The Bible was not far from their minds. If we ignore a few predecessors very determined to demonstrate the truths of Holy Scripture, Cuvier (1769–1833) – even if his religious inclinations cannot be completely forgotten[45] – may be considered as the scientific founder of catastrophism. He has the merit of demonstrating the successive disappearances of animal and plant species; for him, this could only be explained by great upheavals. He evoked several other floods over a period of 80,000 years, which preceded the Flood of Noah's ark and were responsible for the disappearance of many species.

Uniformitarians

From a methodological point of view, the uniformitarians rejected the role of a *deus ex machina* attributed to catastrophes, that is, to unknown phenomena, not proven but opportunely introduced to explain the present situation. They considered that the forces acting today were those that had fashioned Earth in the past. Since erosion, for example, acts very slowly, it is understandable why they were generally believers in long timeframes. To the great forces and short times invoked by the catastrophists, they thus opposed gradual changes and long times. For James Hutton (1726–97), there was an equilibrium (perpetual? Hutton was not explicit on this subject) between the erosion that wears down mountains and the sedimentation that recreates them. As far as Earth's age is concerned, *The Theory of the Earth* begins with a sacred reference, perhaps a little bit tangled:

> that we are enabled to understand the constitution of this earth as a thing formed by design. We shall thus also be led to acknowledge an order, not unworthy of Divine wisdom, in a subject which, in another view, has appeared as the work of chance, or as absolute disorder and confusion.[46]

There follows the acknowledgement of fossils as proof of time passed:

> The Mosaic history places this beginning of man at no great distance; and there has not been found, in natural history, any document by which a high antiquity might be attributed to the human race. But this is not the case with regard to the inferior species of animals, particularly those which inhabit the ocean and its shores. We find in natural history monuments which prove that those animals had long existed; and we thus procure a measure for the computation of a period of time extremely remote, though far from being precisely ascertained.

We then have the emblematic formula of the uniformitarians:

> The result, therefore, of our present inquiry is, that we find no vestige of a beginning, – no prospect of an end.

Finally, at the end of the book, Hutton explicitly sums up his conception of Earth and implicitly also that of religion:

> We are thus led to see a circulation in the matter of this globe, and a system of beautiful economy in the works of nature. This earth, like the body of an animal, is wasted at the same time that it is repaired. It has a state of growth and augmentation; it has another state, which is that of diminution and decay. This world is thus destroyed in one part, but it is renewed in another [...] It is only in science that any question concerning the origin and end of things is formed; and it is in science only that the resolution of those questions is to be attained.

Charles Lyell (1797–1875), who systematized this approach, had a great influence on Darwin. In this respect, we should distinguish methodological uniformitarianism, which postulates the permanence of the laws of physics,[47] and which it would be more appropriate to call 'actualism', from a substantial uniformitarianism that is more debatable and also asserts a constant intensity of the forces in play.[48]

And now?

Earth's real age was to offer the 'uniformitarians' more justification than their opponents, and yet the present state of our planet would be incomprehensible without the effect of 'catastrophes'. Let us take a very popular example, that of the disappearance of the dinosaurs[49] some 65 million years ago (65 Ma). In 1980, the physicist Alvarez noted the fact that a level of iridium one hundred times greater than the average had been discovered across the globe in strata all dated 65 Ma. Iridium is a very rare metal on the surface of Earth, yet abundantly present in meteorites. Alvarez accordingly proposed the following scenario. The disappearance of the dinosaurs was caused by the impact of a meteorite some 10 kms in diameter. The energy of this impact would have been of the order of several thousand hydrogen bombs, causing gigantic fires and ejecting clouds of dust that would have obscured the Sun, creating a protracted 'nuclear winter'.[50] The colossal disturbances caused to the climate, and thus affecting the whole food chain, favoured warm-blooded animals,[51] these being better adapted to sudden variations in temperature and often sea-dwelling or underground, so better protected. The point of impact was subsequently identified as the Chicxulub crater in the Gulf of Mexico.

This thesis has generally supplanted that of a gigantic volcanic eruption in the Deccan plateau in India, which could also have generated a nuclear winter around the same time, as well as leading to the appearance of iridium. The argument, however, is still inconclusive.

It is almost possible to say today that uniformitarianism implies catastrophism: with 'deep times', those of geology, the occurrence of events that are highly improbable and hence extremely rare on smaller timescales, such as large meteor strikes or gigantic volcanic eruptions, becomes very likely.

Darwin versus Kelvin

Charles Robert Darwin (1809–82), founder of the modern theory of the evolution of species, was also highly influenced by Thomas Malthus (1766–1834). We have seen how Darwin, working as a geologist, had recognized an age for Earth of at least 300 million years in the wake of his study of the erosion of the hills of the Weald. The same Darwin, as palaeontologist, despite accepting this ill-founded estimate, would none the less require more time, but without giving figures. The expression 'a vast amount of time' often recurs in his letters, as well as in his major work *The Evolution of Species*. The geological strata several kilometres thick had needed immense time to sediment. It was necessary then to explain:

> why we do not find [in the geological record] closely graduated varieties between the allied species which lived at its commencement and at its close … Although each formation may mark a very long lapse of years, each perhaps is short compared with the period requisite to change one species into another.[52]

This is the famous problem of 'missing links', which led Darwin to reject Kelvin's figures for Earth's age, despite being unable to propose any definitive alternative. His argument can be summed up as follows: since the geological strata do not generally contain intermediate links, and since such links must nevertheless exist, the times of sedimentation of these geological strata, already colossal – certainly tens of millions of years, if not hundreds – are small in comparison with the time needed to account for the actual evolution of flora and fauna. In other words, 50 million years is neither here nor there in relation to geological time.[53] If we add that exceptional conditions are needed to ensure a lasting fossilization, it is understandable why the traces left in geological history

Fig. 6. 'A venerable orang-utan'. Caricature of Charles Darwin as an ape. Published in the satirical magazine *The Hornet* in 1871.

are not significant: they represent only a negligible fraction of the past. The chance discovery in 1861 of a link between dinosaur and bird, the archaeopteryx, possessing both feathers and teeth, was a great support for Darwin. Similarly, in the lineage of our distant ancestors, only *one* Lucy has been found (in 1974) and *one* Toumaï man (in 2001). For the moment, we have only two individuals, a derisory representative of populations that must have counted several hundred thousand individuals (no one knows exactly how many) over millions of years.

His fingers burned by his study of the Weald, Darwin refrained from giving figures, but he viewed Kelvin's range as too short to be credible. As a theorist of biological evolution he suspected a mistake in the great physicist's procedure, but where exactly? He suggested that the Sun might be able to transfer magnetic energy which heated Earth.[54] This was not a good hypothesis, but the idea of a different source of heat was formulated; Kelvin should not be left a monopoly on physics,[55] a man who suspected geologists and other biologists of not accepting the universality of its laws.

Kelvin's authority[56] at this time was so great, however, that his estimate of Earth's age was accepted. The majority of geologists, whose culture tended to be uniformitarian, bent themselves into all kinds of contortions so as to compress their estimates, with the justification that a hotter Earth would have made possible a more rapid evolution of chemical reactions and material actions.

This antagonism between Darwin and Kelvin was abundantly used by Darwin's adversaries. It was not so much Darwin's datings that they were concerned to ruin (those of Kelvin, though shorter, were no more compatible with a literal reading of the Bible), but rather his theory of evolution, which was viewed as an abomination, supposedly having man descend from the ape. It reduced the story of Adam and Eve, as well as original sin, to the status of myth. Russell gleefully cites various quotations from Anglican bishops attacking Darwin.[57] But the situation within the institution was rather more subtle.[58] A crisis was already brewing in the Church of England before Darwin came along, with less literalist tendencies (such as the 'broad church') already developed. Though Galileo had no right to a tomb, or Bruno to a statue, Darwin had the exceptional honour of a burial alongside Newton in Westminster Abbey.

One might have hoped that this polemic from over a century ago would today be closed, as even Pope John-Paul II declared to the Pontifical Academy of Sciences in 1996 that Darwin's theory was 'more than an hypothesis'.[59] Yet this is not the case. The weight of creationists in the United States is such that George Bush, in the course of his electoral campaign, declared, according to the *New York Times* of 22 October 2000: 'I think that, for example, on the issue of evolution the verdict is still out on how God created the Earth.' No further comment is needed.[60]

One might say that Kelvin was the predecessor of today's 'intelligent design', the short timescale he defended being insufficient for the development of life without a preconceived (divine?) plan. But that would be an anachronism: the world of the mid nineteenth century was not that of the twenty-first!

The Death of Earth

Once the concept of Earth's origin is posited, it is quite coherent to predict its death. If we leave aside the Apocalypse and the Last Judgement, there are several scientific hypotheses.

- *Thermic death.* This idea derives from the findings of nineteenth-century physics. Earth is inexorably cooling as a function of the laws of thermodynamics. The planet will not die, but its inhabitants will perish from cold. See the terrifying Figure 7 and its accompanying caption, both taken from an *Astronomie populaire*.[61]
- *The Sun as red giant.* In the Sun's cycle of thermonuclear combustion, a moment will come (in 4 billion years) when it will undergo a tremendous explosion, expanding to beyond the size of Earth's orbit. Earth's inhabitants will be burned to death.
- *Change in Earth's orbit (chaos).* On everyday timescales of hundreds of thousands of years, Earth's movement around the Sun is the very symbol of regularity; it was even used for a long while to define the legal unit of time. But on timescales of a billion years or more, the theory of chaos (see p. 122, below) teaches us that the existence of other planets and the Moon makes prediction hazardous.[62] Earth could be projected either very far from the Sun, or very near.
- *Collision with a heavenly body.* Though highly improbable on a short timescale, this becomes likely in the long term. We should not forget that the Moon was born in the same way more than 4 billion years ago, from a collision between Earth and an object the size of Mars, which has even been given a name: Theïa.

Fig. 7. 'Overcome by cold, the last human family is touched by the finger of death. Their bones will be buried beneath the shroud of eternal ice.' A view of the end of the world by Camille Flammarion (1881).

CHAPTER 3

The Twentieth Century and Radioactivity

The discovery of radioactivity, which marked the twentieth century (and will very likely mark those that follow), engendered an unexpected by-product that would have been quite utopian in the previous century: a clock able to provide an absolute age for Earth, which is today fixed at 4.55 billion years. Better still, it explains how this new phenomenon refutes the calculations of nineteenth-century physicists. Darwin's 'vast amount of time' here acquires its full scientific validity. This is the story, with its twofold implications.

Nuclear Transformations as a Source of Heat

In the late nineteenth century, the Becquerel family had been professors at the Paris Museum of Natural History for three generations. Through Poincaré, Henri Becquerel, its youngest representative, became acquainted with the recent discovery of X-rays by Roentgen in Germany, and he had the idea of revisiting the subject in the light of the work that the Becquerels had conducted on fluorescence. He undertook this study and wondered whether uranium salts exposed to sunlight could subsequently make an impression on photographic plates. Following a routine check,[1] he found that they have this property *even without being exposed to the Sun*: a constant radiation spontaneously emanates from uranium salts. He explored this subject for a while, then abandoned it. It was subsequently taken up by Marie Skłodowska-Curie and Pierre Curie. Discoveries followed one another in rapid succession. The Curies isolated a new chemical element, radium, in 1898, a million times more active than uranium. Scarcely discovered, this metal (the dearest in the world!) became a panacea: advertisements for waters and creams boasted of their radioactivity (Figure 9). In 1903, Pierre Curie in Paris, and Ernest Rutherford and Frédérick Soddy in Montréal (McGill University), demonstrated the release of

Fig. 9. Advertisements vaunting the strong radioactivity of a beauty cream in the 1930s and a soft drink in the 1950s. Hard to imagine today!

heat associated with the disintegration of radium. A radioactive nucleus of radium emits very high-speed particles (alpha radiation). Their collisions transmit their energy to the atoms and molecules in the surrounding environment, which consequently heat up. These particles, accumulating together, form an inert gas that Rutherford and Soddy identified as helium[2] (the particles comprising its nucleus, as we shall shortly explain). The emitting nucleus had been transmuted into a different element.[3]

Shortly after, John Joly (inventor of dating by ocean salinity) had the idea that radioactive elements, present everywhere in the ground, constituted a source of heat that Kelvin had not known about, and that invalidated his calculations. Earth could no longer be viewed simply as a sphere that was initially hot but inexorably cooled by losing energy from its surface, as in Fourier's theory. It contained within it a veritable 'nuclear power station'. Perhaps, it was believed for a while, this could also be the source of the Sun's energy, this being rich in helium? At all events, a radical revision was on the agenda.

It was Rutherford who succeeded in this, in 1904. Here is the account that he himself gave:

> I came into the room, which was half dark, and presently spotted Lord Kelvin in the audience and realised that I was in for trouble at the last part of my speech dealing with the age of the earth, where my views conflicted with his. To my relief, Kelvin fell fast asleep, but as I came to the important point, I saw

> the old bird sit up, open an eye and cock a baleful glance at me! Then a sudden inspiration came, and I said Lord Kelvin had limited the age of the earth, provided no new source (of energy) was discovered. That prophetic utterance refers to what we are now considering tonight, radium! Behold! the old boy beamed upon me.[4]

It is perhaps this anecdote that popularized Kelvin's error.[5] But another of Kelvin's hypotheses was also incorrect, as one of his students, John Perry, had already noted in 1895: Earth's *rigidity*.[6] This hypothesis was important, as it limited the transfer of heat from the interior to the surface simply to thermic conduction, in the way that heat spreads along a rigid metal bar that has been heated at one end. In actual fact, over geological time Earth's interior behaves like a fluid. It is rather like a glacier on a far greater scale: rigid in the short term, but flowing in the longer term; and like rivers, it can hollow out valleys. And, since Earth is not rigid in the long term, heat also spreads by convection, i.e., by transport of matter between the internal hot regions and the external colder ones. Convection is a process of thermic mixing that is far more efficient than conduction: by stirring soup with a spoon, you make sure that the liquid circulates and all parts are equally affected. Its developmental equation is not that given by Fourier, which deals only with rigid bodies without any transport of materials. Earth could then be much older, as Perry had shown.

One may wonder how, paradoxically, a more effective cooling mechanism than conduction makes it possible to obtain an older age. This is because Earth's age had been determined on the basis of the present temperature gradient at its surface (the only one measurable), interpreted in terms of the theory of the cooling of a homogeneous and rigid sphere. Yet Earth as a thermic mechanism is qualitatively more complex than the idealization assumed by Kelvin. To simplify, it is only in the neighbourhood of the terrestrial crust that conveyance of heat is governed by Fourier's equation as used by Kelvin. Deeper down, in the mantle, there is a mechanism of convection that ensures a certain harmonization of the temperature and maintains a high temperate gradient at the surface. How does this happen? Rather than a cannonball, Earth should rather be compared with a cooling cup of coffee.[7] The liquid very rapidly reaches a homogeneous temperature, which then slowly falls by exchange of heat with the porcelain of the cup, itself in contact with the outside air. In space, however, with the brief passage through the

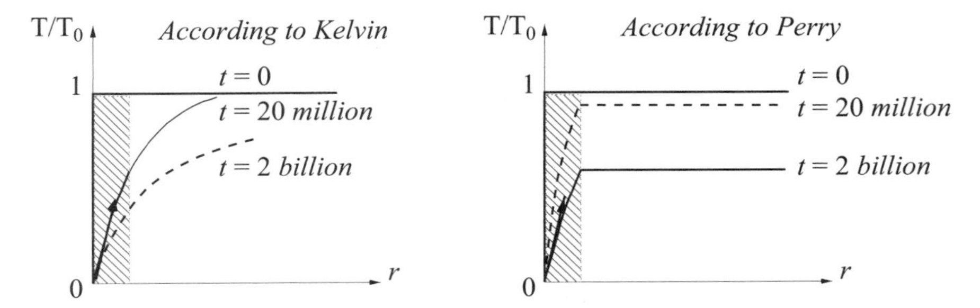

Fig. 10. Diagrammatic temperature curves for Earth T/T_0 as a function of depth r at t = 0, t = 20 million years and t = 2 billion years according to the respective theories of Kelvin and Perry. T_0 is the starting temperature. The grey zone represents the terrestrial crust, and the arrow at the origin the average measured temperature gradient. We see how the two theories, using the same experimental data, i.e., the same temperature gradient, yield completely different ages.

porcelain, the temperature falls sharply, creating a locally high gradient. Figure 10 sums up the respective results of Kelvin and Perry for Earth's age. Earth's interior (the mantle) would be like the coffee, and the thin layer of the lithosphere (the crust) the cup. Perry offered a simplified calculation with two layers, and arrived at an age of 2 billion years.

Thus two factors not taken into account by Kelvin are able to explain the temperature gradient measured at the surface: radioactivity and convection. The second factor is today seen as dominant. As far as Earth's age is concerned, in fact, the discovery of radioactive heat is principally of historic interest: it was simply the first battering ram that ruined the mathematical construction that Kelvin had bitterly defended.

We should remember that the upper limit set by Kelvin in his estimate of Earth's age – 40 million years – followed from his estimate of the age of the Sun, the two bodies being rightly viewed as contemporary. Here again, though, Kelvin was unaware of the physical mechanism that lay behind solar energy. The best source imaginable in his day was gravitational collapse, which does indeed play the driving role until the temperature at the centre reaches around 15 million degrees. But then – unknown to Kelvin – nuclear fusion reactions take over, which explains both the Sun's equilibrium and its production of energy for billions of years (see Appendix C).

Radioactivity as an Absolute Clock

Very fortunately, radioactive disintegration would also provide a method for the absolute dating of rocks. In 1902, Rutherford and Soddy established the notion of the period[8] of a radioactive element (see Appendix C). To characterize the radioactivity of an element, they measured its activity, i.e., the number of disintegrations it undergoes in a unit of time. They noticed that the time needed for the activity of a radioactive element to fall by half was a constant that depended only on the particular nucleus.[9] This time, called the T period of the radioactive element, is characteristic because it is also independent of the initial quantity of radioactive nuclei: at each interval of time T, the number of radioactive nuclei remaining divides by two. It is this regularity that makes radioactivity a usable clock; it has the consequence that radioactivity decreases exponentially. We owe it to Rutherford to have made use of this law of decrease for the dating of rocks, which enabled him to obtain a lower limit for Earth's age.

To understand this method, suppose:

- that a stable (non-radioactive) element, which we call for example Pb, contained in the sample of rock, can only be obtained as the product of the disintegration of a radioactive element, which we call U, initially present in the same sample;
- that this sample has remained *closed*, i.e., without any exchange with the outside;
- that the period of U is known, say for example 5 billion years.

If today we find a number of atoms of Pb equal to those of U, this means that the initial quantity of U has fallen by half, since an atom of Pb can only arise from the disintegration of an atom of U. The conclusion is that the sample has a date of 5 billion years, the period of element U. To avoid any calculation here, we have taken the simplest case, with exactly half of the initial nuclei having disintegrated. But it is very easy, with the aid of the law of exponential decrease, to find the age of the sample once the proportion of atoms Pb/U is known by experiment.

Presenting the principle of the method, however, shows the difficulties it involves. For example:

- In the case of uranium which disintegrates into lead, the connection is not direct. Uranium already exists in the natural state in two isotopic forms, U-235 and U-238. The first leads to the lead isotope Pb-207 after a chain of disintegration involving 11 unstable nuclei and the emission of 7 alpha particles (helium nuclei); the second leads to a nucleus of Pb-206 after a chain of 14 unstable nuclei and the emission of 8 alpha particles.

- A certain quantity of lead, N_0, may already have been present at the time that the rock solidified and became a closed system. This has to be known *a priori* in order to directly calculate the age of the sample on the basis of its radioactive decline. But how? We shall see below (p. 42) that it is possible to know N_0 for the case of lead, and in Appendix C how a clever trick makes it possible to dispense with this in the case of dating by rubidium.
- Checking the closure of the rock may be difficult, especially if the products of disintegration are gaseous.
- The determination of the period of an element can be a delicate matter, especially when it is in the order of billions of years, since this means there are very few disintegrations per second.
- The age arrived at is that of the sample at the date of its closure; but a model is needed to deduce Earth's age from this.
- Finally, measurement of extremely weak concentrations of isotopes requires very precise instruments. These were developed particularly in connection with the manufacture of atomic weapons.

In order to date rocks, Rutherford initially used the quantity of helium given off by the radioactive disintegration of uranium. He obtained an estimate of 497 million years, which he knew to be an underestimate, since a large quantity of helium had certainly escaped from the rock.[10]

In 1905, B. B. Boltwood realized that as well as helium, the disintegration of uranium also produced lead. In fact, basing himself on the geological ages of rocks, he established that the Pb/U relationship was higher in proportion to the age of the rock. He therefore proposed a more reliable method by measuring lead as the final (and non-volatile) element of the disintegration of uranium (see Appendix C for a detailed modern description of this kind of method). This enabled Boltwood to obtain in 1907 an age spread between 410 and 2,200 million years.[11]

The age limit imposed by Kelvin was now abandoned: his hypotheses were not justified, and his conclusions were contradicted by the new findings.

Different methods

Today, dating generally uses several different disintegration chains: potassium—argon, thorium—lead, uranium—lead, rubidium—strontium. In the majority of cases, it is possible to test the internal coherence of the model by verifying that the results of measurement lie on a theoretical curve (see the example of the straight isochrones in Appendix C); these are then compared with the results obtained independently on other nuclear chains. We shall not go into the different techniques used here. Let us simply say that this process has developed in close symbiosis with the advances of atomic and nuclear physics, including, among others:

- 1911, the discovery of the atomic nucleus by Rutherford;
- 1912, the discovery of isotopes by Soddy makes it possible to understand the disintegration chains;
- 1932, the discovery of neutrons by Chadwick explains the existence of isotopes of a single element;
- 1939, Hans Bethe's explanation of the cycles of thermonuclear reaction inside the Sun (see Appendix C) makes it possible – at last – to understand the origin of the energy it emits.[12]

Other methods, more rarely used, rely on rocks acting rather like photographic plates, with fixing times of millions of years. They either exploit traces left by the spontaneous fission of uranium-238, or those of alpha particles disintegrating. These traces, which are stronger in proportion to the fixing time, give a way in to the age of the sample's formation. Account must be taken of the fact that any heating of the rock partly destroys this signature.

Rocks as photographic plates

1. *Radio halos.* The alpha particles emitted by a radioactive source trapped in black mica (biotite) can damage the crystalline structure surrounding it and in this way create a spherical halo visible to the microscope. Its radius is a function of their average kinetic energy, thus of the nature of the source. The colour, after standardization, gives the time of irradiation and, on this basis, the age of the sample after its crystallization. The idea goes back to Joly (once again!) in 1907.

2. *Spontaneous splitting.* In 1940, it was realized that certain elements, such as uranium-238, can split spontaneously, i.e., divide into two adjacent fragments of mass.[13] The products of this split destroy the crystalline network in their trajectory. The minuscule traces thus produced are enlarged by appropriate acid treatment, then counted under the microscope. Their number makes it possible to estimate the total production time of the fragments of splitting once the probability of this event is known.

The list of modern researchers here is very long. The two that stand out are Arthur Holmes (1890–1965) and Fritz Houtermans[14] (1903–66), who independently proposed the 'isochrones' method in 1946 (see Appendix C). This method made it possible to set the lower limit for Earth's age back to 3 billion years.

An Earth older than the universe?

In the 1930s, however, the 'origins' of the universe were dated by astronomers at 2 billion years. This was the age estimated on the basis of Hubble's determination of the universe's rate of expansion.

By interpreting the difference in frequency of radiation originating in distant galaxies as a Doppler effect,[15] Hubble demonstrated a general expansion movement of the universe. He also showed that the more distant a galaxy is from us, the greater the speed at which it moves away.[16] This general movement of matter on the large scale is characterized by a constant, known as the Hubble constant, that expresses the increase in speed per megaparsec[17] of increase in distance. Hubble calculated a value for this in the region of 500 km/s/megaparsec. This Hubble constant, which is unaffected by time, makes it possible to put a date to the Big Bang.

The improvement in experimental techniques, in particular the measurement of distances, led to a significant reduction in the Hubble constant, by a factor of seven, and consequently a corresponding increase in the age of the observable universe. The value now accepted is 72 km/s/megaparsec, corresponding to an age of some 13.7 billion years. The age of the solar system is thus no longer in contradiction with cosmological results.

Finally, the work of Clair Patterson (1922–95)[18] did much to refine these results.[19] He understood that the oldest accessible rocks, by definition situated on Earth's surface,[20] can only provide the age of the consolidated terrestrial crust: this is permanently renewed by movements of subduction (in which one tectonic plate slides under another) and by volcanic fractures.

Where then can genuinely primitive samples be found? And primitive in relation to what?

The age of Earth deduced from that of meteorites

The only rocks available are naturally located on Earth's surface, and dating methods only give the date of their *closure*. Earth is necessarily older than this. It is only the oldest meteorites that can offer a closure date contemporary with the formation of the solar system. To move from the age of these meteorites to that of Earth presupposes a model of how the planetary system was formed.

The formation process of a star and its system of planets includes different episodes of varying duration.[21] It all begins, within a cold molecular cloud, by the gravitational collapse on itself of a cloud of gas and dust. The initial cause that triggers this remains debated.

Certain models envisage[22] the random appearance of phenomena of this kind in the cold zones of molecular clouds (T<200°K); others assume that this collapse is directly triggered by a shock wave produced by the explosion of a massive star at the end of its life (supernova). In the latter case, the explosion of the massive star must take place in the nearby environment (a few dozen parsecs) of the future planetary system, and in a reasonable timeframe (a few million years at most before the gravitational collapse). It is then possible to envisage the proto-solar cloud, i.e., that portion of the molecular cloud that will be isolated to form the solar system, being enriched by new nuclei that this supernova has produced. These are so-called heavy elements, 'trans-ferric' ones such as lead or uranium, which are not produced by nuclear reactions within normal stars such as the Sun. Before the central proto-star reaches a mass sufficient to initiate the thermonuclear reactions of hydrogen combustion, we can distinguish schematically two phases whose average durations are different. There is the so-called embedded phase, which in the case of a star of the Sun's type lasts around 200,000 years. During this phase, the cloud of gas and dust still completely surrounds the proto-star and the rate of accretion is very high, the proto-star passing from 0.5 to 0.8 of the solar mass. Then comes the 'revealed' phase in which the cloud of circum-stellar matter flattens to form an accretion disc, which will subsequently become the proto-planetary disc. This phase lasts from 3 to 5 million years, until the disc has been completely dissipated. It is estimated that the minimal mass of the disc that finally manages to escape the final accretion into the star, in other words the mass of gas needed to form the present planets, is in the order of 1 to 2 per cent of the solar mass. In our solar system, this mass is overwhelmingly contained in the two giant gas planets, Jupiter and Saturn.

The process of planetary formation still remains marked by much uncertainty. For example, it is very hard to explain how these grains of interstellar dust of micrometric size congregate to form the objects of the size of a kilometre whose movement is capable of freeing itself from that of the gas. A complex process of accretion and destructions (due to collisions) is produced, forming bodies of asteroidal size, then embryonic planets of lunar size and finally the planets. It is generally held that the whole process that forms a planet like Earth lasted more than a hundred million years. And so the rocks most contemporary with the formation of the Sun are to be found among the smallest objects, the meteorites.[23] We should note that, in any case, the order of uncertainty is low in relation to 4.5 billion years, only a couple of per cent.

Historically, the most precise method for determining the age of these meteorites is the lead–lead method, developed by Clair Patterson. This consists in analysing the composition of lead in stable isotopes of its refractory inclusions. Pb-206 is produced by the disintegration of uranium-238, which has a half-life of 4.51 billion years; Pb-207 comes from uranium-235, whose half-life is 710 million years. Pb-204, on the other hand, is not of radioactive origin. To reach an age from the measurement of isotopic ratios, the initial composition must be known. Fortunately, analysis of the meteor that gave birth 25,000 years ago to Meteor Crater in Arizona showed that it did not contain uranium. Taking into account the half-lives, then, if it does not contain uranium today, it did not contain it originally. The proportions of its lead isotopes are therefore those of the primitive cloud (assumed to be chemically homogeneous), and it can serve as an initial reference point for all meteorites, including those that do contain uranium.

Other disintegration chains have since been studied, notably that of hafnium-182 and iodine-129, thus refining Patterson's results. The most recent dating results of the refractory inclusions of meteorites[24] give a convergent age of 4,568 billion years, with an uncertainty of 700,000 years, i.e., less than 0.02 per cent![25]

It is impossible to claim a similar precision for Earth's age: the difficulty, as we saw, is quantifying the duration of the accretion period. The best determination today is 4.55 billion years, with an uncertainty of around 50 million years. And so the question remains somewhat subjective: in the course of the complex formation process, what are the criteria by which we recognize the body in solar orbit as Earth, and thus at what moment in time?

In the nineteenth century, calculation of Earth's age did not proceed in a steady fashion from the false to the true. Even if there was never a consensus, at the start of the century the champions of a biblical Flood – by now a little aged – were still numerous; it seems however that by 1840, the majority of geologists accepted Lyell's ideas, and thus an immensely long age for Earth.[26] Then came the 'help' of physics in the person of Lord Kelvin, which reduced this infinity to 20 million years. That was far too short for Darwin, being incompatible with his theory of evolution. The discovery of radioactivity, at the end of the century, made it possible both to invalidate Kelvin's theory and to attribute an absolute age of 4.55 billion years to Earth. In the end, the physicists proved the naturalists correct.

Really 4.55 billion years?

After this long history of the evolving age of our planet, a question is raised: are we not just as naïve as Ussher, Newton or Kelvin, who also believed that they possessed the 'true' answer? Why should these colossal variations in the age estimates stop? Why do we believe that, though there has been an evolution in the calculation of Earth's age, this will not continue in the future?

Before showing why we believe that we possess the 'true' answer, or, to put this in a seemingly more rigorous way, why it is extremely probable that this is the case, let us draw three lessons from this long story:

1. There was first of all a reversal of intellectual priorities. We showed how, until the Renaissance, the conception was that the truth about Earth's identity (its age, but also its position) was

contained in the Bible; the nascent science of geology was supposed to justify the Holy Scriptures to the letter. Now it is science that compels a 'correct' interpretation of the biblical Genesis. Today, except for a few creationists, it is commonly accepted that the Bible is not the source of scientific knowledge or even historical knowledge. We should note the good faith – in every sense of the term – of the scientists who challenged biblical dating: as with Earth's movements, their starting point was always the sacred text that they sought to verify.

2. Kelvin was the embodiment of scientific procedure, which did not prevent him from being mistaken. However, as he did not base himself on revealed and thus 'undebatable' truths, but on verifiable hypotheses (which he wrongly thought were verified), challenge and hence progress was possible.
3. Earth was accordingly dated first at 4,000 years old, then at 400 million, then at 20 million years. Are we justified today in asserting that 4.55 billion years is the truth? Clearly, this age may wobble a bit; at all events, the uncertainty about what one may call Earth's date of birth still has an irreducible margin of some millions of years. But today, all approaches (astronomy, geology, physics, biology) are in agreement. We shall never go back to the transition from the biblical 4004 BCE to the several tens of millions already effected in the nineteenth century, then to the billions imposed by radioactivity. To be sure, certain hypotheses may prove false. The accretion model of Earth following very closely on the formation of the Sun is not immune from small modifications. The periodicity of radioactive elements is supposed to be constant; if it had varied over time, Earth's age would be modified accordingly. Since no experiment nor any theory up till now suggests this, it is reasonable to suppose that any such variations that might be shown to have occurred would be minimal and with little effect.

There is no direct measurement for Earth's age (in fact, there are in general few direct measurements in physics), but each time, as we have shown, a build-up of experimental results and their agreement with theories or models.

For Ussher or Newton, theory meant the Bible, the divine source of truth. Empirical results were obtained not from experiment, but from

Holy Writ. No confrontation with experimental results was envisaged. The procedure might often be scholarly, indeed very much so, but it was scholastic, that is, *founded on texts.*

With Kelvin's absolute dating we embark on a scientific approach. This can be illustrated by a simple yet typical example: determination of the age of a tree, in the absence of a human witness.

- There is first of all a theory, dendrochronology; each year a new ring appears, this phenomenon being understood as a physiological effect due to the seasons.[27]
- The experimental results here are the rings of the tree sections. A count of these determines the age. Is this unchallengeable? No. As in every scientific field, the result can be challenged. How? For example, because it assumes that all the rings are visible and that no other phenomenon is responsible for them, because it also assumes that the duration of the seasons has never varied, etc. But listing its weaknesses simply shows the strength of the theory.[28]

Kelvin's theory must be analysed in the same manner. Even in its own time, however, it was not so solid:

- It contradicted Darwin's theory of evolution;
- It accounted for the empirical results of geologists only at the price of rather unconvincing juggling;
- Perry presented objections regarding the role of convection;
- Chamberlin proposed other possible sources of heat.

The age of 4.55 billion years for Earth, on the other hand, resting on the theory of universal radioactive decay, is a conclusion with as solid foundations as, for example, dendrochronology. The age now given is the result of converging independent measurements, and does not at all contradict any empirical or theoretical result, whether in astronomy, geology or palaeontology. It is therefore what we call a scientific truth. But it is *only* a scientific truth:[29] for the time being, we do not know any others that are 'more' certain.

By comparison, the estimated date of the Big Bang, in the order of 14 billion years, rests on theoretical and empirical premises that are still hotly debated, and is what we might call a 'far less certain truth'.

If Earth's age clearly does not lend itself to direct experimental measurement,[30] it does result, by way of radioactivity, from a mass of measured results that are independent and repeatable. In this sense – as is the case in many domains of science – it is in the final analysis established experimentally.

The importance of a correct estimate of Earth's age is considerable:

This age of 4.55 billion years is not just one more figure in the series of ages, of interest only to astrophysicists; only this timescale renders intelligible the establishment of the order of the solar system and the fantastic complexity of life.

Part Two

EARTH'S MOVEMENT

> Science replaces the complicated visible by the simple invisible.
>
> – Jean Perrin, *Les Atomes* (1913)

> The method of rising from the abstract to the concrete is only the way in which thought appropriates the concrete, reproduces it as the concrete in the mind.
>
> – Karl Marx, 'Introduction to the Critique of Political Economy' (1857)

We[1] are going to trace the history of the ideas that led to what everyone learns at school today: *it is Earth that turns around the Sun, and not the Sun that turns around Earth*. But, as we are also taught at school, there is only relative movement, the movement of one object in relation to another; it makes no sense to speak of absolute movement. Is this a contradiction?

Everyone has had the experience of being in a train at a station that starts moving without any shock. With your mind caught up in a newspaper, you suddenly lose your bearings when you notice out of the corner of your eye that the station, the whole station, along with the other trains and busy passengers, has silently begun to move. A fraction of a second is needed, that is, an interval of time accessible to consciousness, for reason to regain the upper hand and reassure us: stations do not move, but trains do. This is what a locomotive is for; and so it is our train that has started moving. In other words, we bring to the analysis of the situation a knowledge *external* to our immediate experience.[2]

Now, in the case of Earth and the Sun, neither of the two bodies is equipped with a driving mechanism that explicitly ensures its movement and would thus enable us to decide without hesitation. And so, if there is only relative movement, why privilege the movement of Earth rather than that of the Sun? We could decide to adopt *the point of view of someone on Earth*, motionless by definition, who looks at the sky and tries to decipher the movements of the stars. Why not just say *what we see*, that is, that the Sun turns around Earth? Several studies indicate that a significant fraction of the population (around 20 per cent) believe that this is indeed the case. Should we conclude that what science teaches is wrong, and prefer what our senses indicate? Conversely, are we sure that what is given as the correct response has dealt with the objection of the relativity of movement? Have we not accepted too rapidly the argument of authority, which compels us to ignore our own perception of things?

A very strong motivation is needed to privilege the current view, that of focusing our thought on a fixed Sun and examining the movement of Earth from it.

This second part of the book deals with the slow emergence of this motivation. It did not take shape as a wide avenue leading straight from ignorance to knowledge. In this urban metaphor, rather than an American-style city it is the structure of a medieval town, with its interlacing alleys, detours, crannies and dead ends, that is more appropriate to represent the sometimes erratic movement of science, including that of the astronomy of the solar system. Significant advances were made suddenly, after long periods of maturation, stagnation, even regression. We shall see how the very particular relations between physics, mathematics, observations and measuring instruments became the hallmark of modern physics.

The decentring of Earth, the famous Copernican revolution with its profound and multifarious consequences, was the result of a combination of three factors: 1) the conviction that an overall geometrical representation of the universe, based on what would today be called principles of symmetry, is possible; 2) an increasing precision of measurements; 3) a mathematizing of reality, following the rule of systematic comparison between the predictions of models and empirical results.

The precision of astronomic observations in the ancient world was often remarkable. But it is to the Greek scientists, and them alone, that we owe the idea of placing these studies in a geometrical framework. This was an intellectual act of amazing boldness: the invention of a celestial geometry, projecting onto the skies the conceptual instruments – straight lines, circles, parallels, proportions – invented to measure the areas of fields!

The stumbling block of these models was the phenomenon of retrogradation of the planets: from time to time, the movements of the planets appear to go into reverse. We shall develop this point in detail, on account of its importance. It was to account for this in a simple way that astronomers of the sixteenth and seventeenth centuries were led to abandon geocentrism and develop various heliocentric models that, from Copernicus through Galileo to Kepler, paved the way for the emergence of Newton's mechanics. From that time, if from the standpoint of the *kinematics* of Earth's movement (its description), the geocentric and heliocentric views were in principle equivalent, they were no longer so from the standpoint of *dynamics* (the representation of causes): here we have the answer to our opening question.

We shall describe the various forms of resistance to the Copernican revolution, illustrated by Galileo's struggle to extricate himself from the Aristotelian straitjacket, not only against the church, but also against common sense.

We shall then deal with the measurement of the distances and dimensions of the observable universe. The celestial geometry of the Greeks had made it possible already in the second century BCE, with an elegance still perceptible today, to obtain the dimensions of Earth, the Moon, the distance of the Moon from Earth, and the distances of the various planets relative to the Earth–Sun distance. They even believed that they had an estimate of Earth's absolute distance from the Sun. But if all the relative distances were correct, it took a further seventeen centuries to realize that the Earth–Sun distance had been underestimated, in the best of cases, by a factor of eighty! Once this was corrected, it was possible to obtain the distances of the nearest stars.

CHAPTER 4

Before Copernicus

Anyone can look at the sky with the naked eye, as the ancients did. But equipped with our modern culture, it is far harder for us to examine the theories and concepts of the ancients with the same naked eye. When, for example, we maintain that for Aristotle speed is proportionate to the force that creates it, this is a summary in modern terms; it was not how the master expressed himself. The separation of chemistry from alchemy, of astronomy from astrology,[1] however commonplace for us, did not exist in the same way before the Renaissance. We need therefore to say a few words about astrology before limiting ourselves to the astronomic part of ancient texts.

The relationship between astrology and astronomy

Every civilization has studied the motions of the heavens, initially for quite practical reasons: to establish the calendars needed for agriculture and religious festivals, as a guide for ships and, last but not least, for astrological predictions proceeding from a unified view of man and the cosmos. In the oldest societies, where men were so exposed to the vagaries of nature, this was omnipresent, threatening or benign, but never neutral or remote. To know how to interpret correctly the advance signs of catastrophe could be vital. Primitive thought saw causalities everywhere, for fear of missing one. 'The first difference between magic and science', wrote Claude Lévi-Strauss, 'is therefore that magic postulates a complete and all-embracing determinism. Science, on the other hand, is based on a distinction between levels: only some of these admit forms of determinism; on others the same forms of determinism are held not to apply.'[2] Astrology is one facet of this primitive, pre-scientific, thinking.

From antiquity to the Renaissance, astrology was an important – perhaps the most important – practical application of astronomy. Copernicus, Tycho Brahe and Kepler all believed in it. For Galileo this is far less certain, yet each of the four certainly drew a substantial income from it. That was not the case with Newton, nor of course for almost any scientist since.

Astrology underwent a slight change in status with the transition from Mesopotamia to ancient Greece and Rome.[3] For the Babylonians, the stars gave indications of the way in which the gods were concerned with the fate of kingdoms; individual futures were the province of local divinities, it was they who had to be approached. The Greeks extended the virtues of astrology to private destinies. The progress of mathematical astronomy had made the erratic movements of the planets calculable, so that they moved from premonitory signs to the causes of future events. That is more or less the origin of modern, so-called scientific, astrology, which sees the fate of an individual as governed by the situation of the stars at the time of his or her birth.

The great monotheistic religions have always condemned astrology on principle: it threatened to encroach on divine power (and, more prosaically, on the power of the clergy). The council of Laodicea (c. 364 CE), whose most well-known decision was to forbid the ordination of women, was the first in a long series that banished astrology. These successive prohibitions show that, in actual fact, the practice of astrology was common. The courts of the Christian kings contained official astrologers; it is even reported that Urban VIII, the pope who condemned Galileo, secretly had Campanella[4] as his personal astrologer. Islamic tradition was perhaps the most hostile to astrology. The following pleasant *hadith* is attributed to the prophet Muhammad: 'The astrologers will have lied, even when they foresaw what actually came to pass.'[5] Besides, Arabic scientists generally distinguished their astronomical works from their astrological ones. This however did not stop viziers and sultans from being often surrounded by astrologers. Finally, if illustrious rabbis also practised this supposed science of the stars, that was not the case with Maimonides, which leads us to assume that it was rejected by the Jewish scholarly milieu.

What Do We See with the Naked Eye?

To measure the great scope of the evolution of ideas, and the obstacles this had to overcome, we must try to put ourselves back in the conditions of the ancients, who could only observe the skies with the naked eye; it was not until the late sixteenth century that Galileo had the idea of aiming at the sky the reading-glass that had come from Holland. That said, eyes were never quite naked; they were aided by instruments for pointing and for measuring angles, with a steadily increasing size and sophistication: we can only imagine the gigantic instrument, 40 metres in diameter, built by Ulugh Beg close to Samarkand in the fifteenth century.

Diurnal motion

This is the easiest to note. It is only necessary to examine the sky at an interval of a few hours to observe that all the stars, while keeping their fixed relative positions, have effected a rotary movement from east to west, on an axis that runs through the centre of Earth and passes (at the present time) very close to the Pole Star. These celestial objects are generally called the *fixed stars*, not because they seem immobile, but because they occupy unchanging positions in relation to one another. Today, long-exposure photography can show perfect circular arcs as the traces of each star's apparent motion, from which one can deduce that the complete rotation is effected in slightly less than 24 hours: what is called the diurnal motion.

In all civilizations, certain groupings of stars have been seen as having particular forms, and been given evocative names, some of which have persisted through to our own day. In the West, the names of several constellations go back to Babylonian times, such as the Plough, the Bull and the Scales.

Annual motion

At any given place and time, the constellations visible at night change in the course of the year. The (sidereal) year is in fact defined as the interval of time separating two identical positions of the constellations in the sky. Its length is a little more than 365 days. What is the Sun's motion against the background of the fixed stars, and how can we observe it, given that, except in the rare cases of eclipses, the light from the Sun makes it impossible to see the stars in daytime? The answer to this was found when it was understood that it is the Sun that illuminates the Moon. When this appears dark (new moon), it is because the Sun, for us terrestrials, is behind it. Conversely, when the Moon is full, as shown

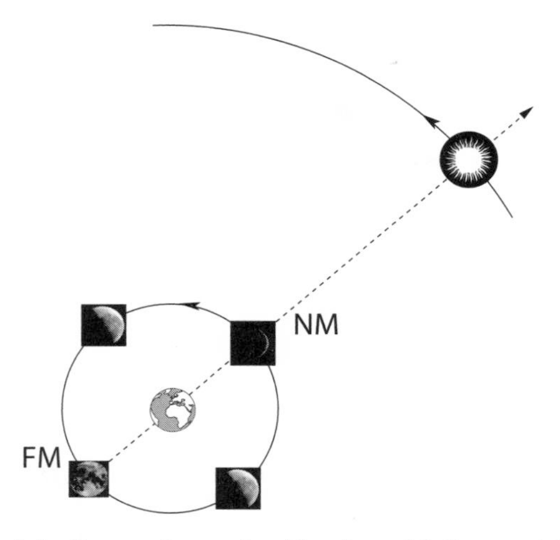

Fig. 11. Position of the Sun as determined by that of full moon FM. NM indicates new moon. Because the Moon's trajectory is not exactly situated in the ecliptic plane, Earth does not hide the Sun at full moon, except in the case of a lunar eclipse.

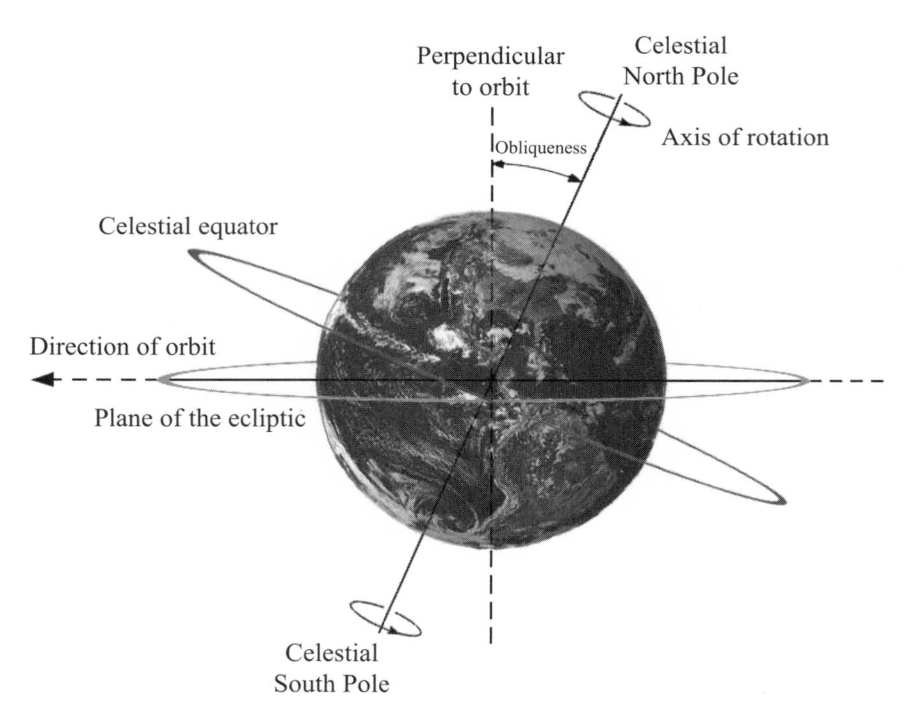

Fig. 12. Inclination of Earth's axis of rotation (an angle of 23°) and its relation to the planes of the ecliptic and the celestial equator. Earth is shown as viewed from the Sun, and the direction of its orbit from this direction is anticlockwise.

in Figure 11, this is because the Sun is in opposition to the Moon. The ancients noted that from one full moon to the next, the Sun moved east in relation to the background of the fixed stars, describing a trajectory known as the ecliptic. This planar and almost circular trajectory makes an oblique angle of approximately 23° with the plane perpendicular to the axis of rotation of the fixed stars (Figure 12).

The retrogradation of the planets

Apart from the Sun and the Moon, five heavenly bodies visible to the naked – and unblinking – eye stand out against the fixed stars. We know them today by their Latin names: Mercury, Venus, Mars, Jupiter and Saturn. (Uranus and Neptune were unknown to the ancients, being too remote to be visible to the naked eye.) As well as their diurnal motion, these planets ('wandering stars' in Greek) move in relation to the fixed stars as the Sun does, that is, eastward, and closely following the same line of the ecliptic. But their movement presents irregularities. From time to time, their eastward march slows, stops, and reverses westward, before stopping again and continuing east. This movement is known as a retrogradation. Understanding the origin of these irregular movements was one of the great challenges of ancient astronomy.

Astrology is naturally very keen on planetary retrogradations. Just think of it! If you are under the influence of a planet, this hesitation in its movement must necessarily have important – even disturbing – consequences on your life ...

How Do We Now Understand the Movements of the Planets?

All the phenomena we have just summarized can be simply explained on the basis of two movements that are well known today:

- On the one hand, the rotation of Earth on itself. This diurnal motion is responsible for the alternation of day and night.
- On the other hand, the movement of Earth (and other planets) around the Sun (Figure 13). This accounts for the seasons, and simply explains the phenomenon of retrogradations.

The seasons

Contrary to a stubborn belief, the reason it is hot in summer is not because the Sun is any closer to Earth.[6] The inclination of 23° between Earth's axis of rotation and the plane of the ecliptic is responsible for the alternation of seasons. In summer, the Sun's rays reach the surface of the globe's northern hemisphere more vertically on average, making for a greater insolation per unit of surface (see the inset on Figure 13); on top of this, the loss due to absorption is less, on account of the shorter thickness of atmosphere it has to penetrate; and finally, the hours of sunshine are longer each day.

Figure 14 illustrates, in the heliocentric representation, the expressions of ancient astronomy (and astrology!), such as 'the Sun is in the constellation of Libra, Aries, etc.'[7]

Retrogradations

Retrogradations are an effect of perspective. Let us take the example of an outer planet (that is, one farther from the Sun than Earth is). The speed of a planet is greater the closer it is to the Sun. Earth, accordingly, moves more quickly in its orbit than Mars does in its orbit, and as a result the Martian year is 1.85 terrestrial years. Being more rapid, and with less distance to travel, Earth thus regularly overtakes Mars in the course of its movement. Each time it does so, we have the impression,

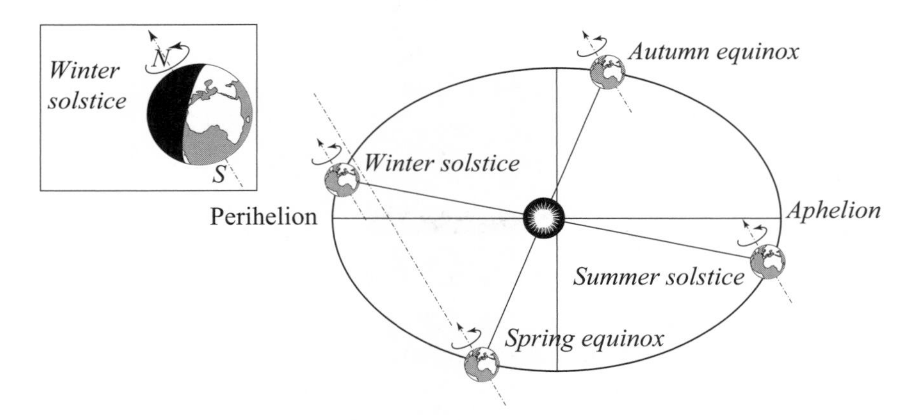

Fig. 13. Heliocentric representation of the solar system. The inset on the left is an enlargement of Earth's position at the winter solstice. We can see how in winter the northern hemisphere N receives less sunlight than the southern hemisphere S.

seen from Earth, that Mars slows down, stops, then for a few weeks regresses against the background of the sky, before resuming its course in the initial direction (Figure 15). As Mars and Earth do not orbit in exactly the same plane, the trajectory of Mars does not follow a straight line along which it retreats, but sometimes appears as an S, other times as a loop.[8]

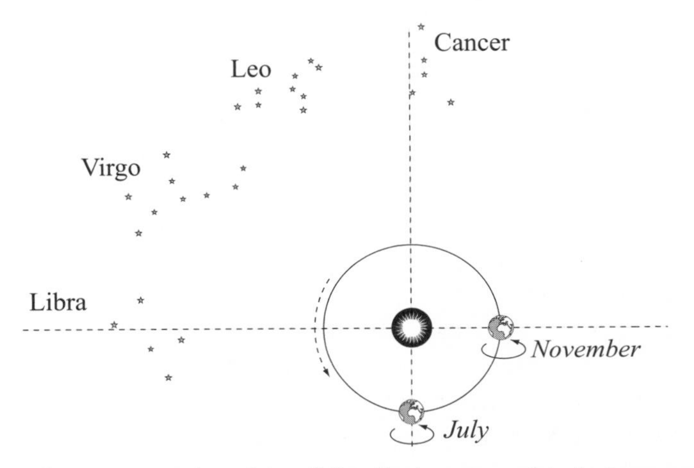

Fig. 14. Heliocentric explanation of the Sun's apparent trajectory against the background sky. In July, the Sun's image is projected on the stars of the Cancer constellation, and in November on those of Libra.

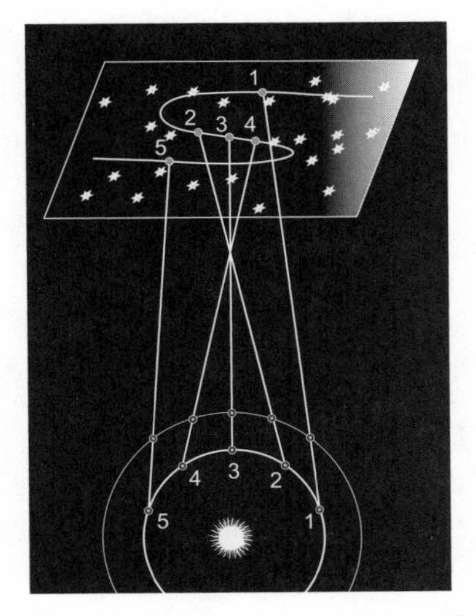

Fig. 15. Heliocentric explanation of the retrogradation of an outer planet. When Earth successively occupies positions 1, 2, 3, 4, 5, we see Mars slowing down and describing an S-shaped curve.

What interval of time separates two retrogradations of a planet?

In the case of Mars, therefore, we observe a retrogradation approximately every two years. To understand this mechanism, let us assume the situation of retrogradation shown in Figure 16: the Sun at S, Earth at E_0 and Mars at M_0 are aligned, with Earth being at its closest to Mars (said to be in conjunction). A year later, Earth (at E_1) has circled once around the Sun, but Mars (at M_1) has only undergone half of its orbit.

Earth and Mars, then at their farthest from one another, are said to be in opposition. Nearly a year must go by until they are once again in conjunction, with therefore a new retrogradation. Now take the case of Jupiter. Its year lasts 11.86 terrestrial years, or 12 at a close approximation. If it were exactly 12, a simple calculation[9] would show that the interval of time separating two retrogradations would be 1 and 1/11 years. Analogously for Saturn, whose time of revolution around the Sun is close to 29 terrestrial years, the retrogradations are separated by 1 and 1/28 years. For the inner planets the same reasoning applies, but switching the roles of Earth and the respective planet. It can be deduced from this that for Mercury, whose year lasts only about one quarter of the terrestrial year, we observe three retrogradations per terrestrial year. And finally for Venus, whose year is close to half the terrestrial year, we observe one retrogradation per year.

Figure 17 shows the trajectories[10] of the Sun and planets *around Earth*, as seen looking down on the ecliptic plane.

But these are still not what we see from Earth, as Earth is situated very close to the same plane. It is necessary then to pivot the page of

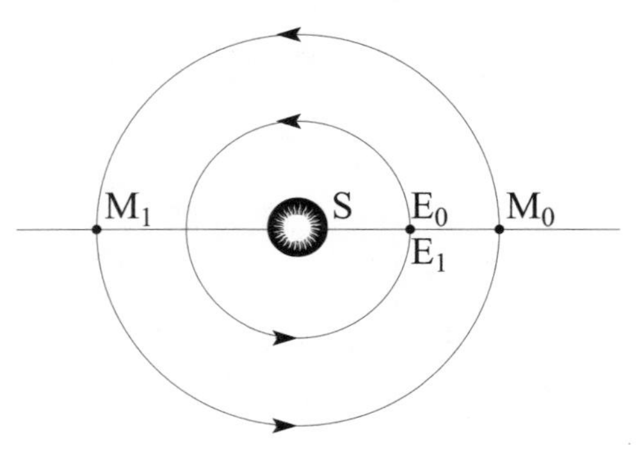

Fig. 16. The rotations of Mars (M) and Earth (E) around the Sun (S). At the initial moment the planets are respectively at M_0 and E_0 (conjunction), and after a terrestrial year at M_1 and E_1 (opposition).

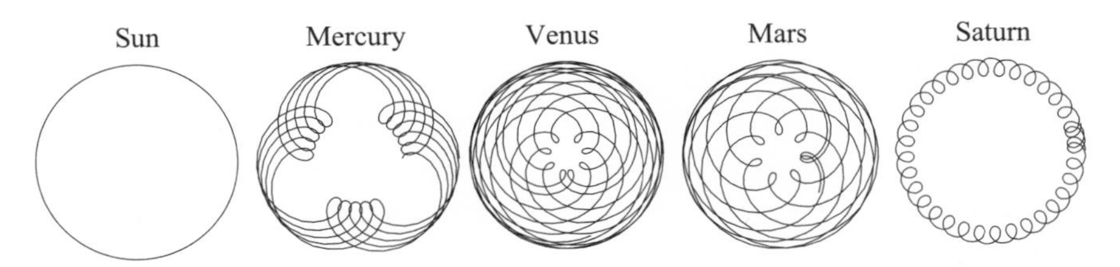

Fig. 17. (Complicated) trajectories of the Sun and five planets in the ecliptic plane, with Earth fixed in the centre. The equations for these curves are given in Appendix D. The orbital times for Mercury, Venus, Mars and Saturn respectively are approximately 5, 8, 16 and 50 years.

the figure and look at it edgewise: in this operation, the perception of distances disappears, and all that remains are points moving along a line, as the planets do on the ecliptic. At certain moments, however, this movement from east to west reverses.

The majority of celestial objects seem to coincide in turning around us from east to west in 24 hours. These are the fixed stars. The Sun also shows an annual eastward movement in relation to these. The planets do the same, but according to periods that are specific to them, and with retrogradations from time to time.

We shall now go back to tell the history of the transition from the 'movement of the heavens' to that of the movement of Earth, as a very fine illustration of Jean Perrin's phrase given as an epigraph to Part Two. To understand the genesis of the movement of the heavens in ancient Greece, the acknowledged cradle of modern science, we have to know that there was a real circulation of knowledge in antiquity between Babylonians, Indians, Greeks and, more marginally, Chinese.[11] It would be the Arabic scientists who rekindled the Greek torch, which was then imported to Europe.

The Contributions of Egyptian and Babylonian Astronomy

Did astronomy, as a science of the observation and understanding of the movement of the heavenly bodies, appear in Egypt, Mesopotamia or Greece? As far as Egypt goes, the answer is no. The contributions of Egyptian scientists to astronomy were very modest, compared with their contributions to medicine and certain fields of applied mathematics. Perhaps this was due to the spirit of this civilization, which was at the same time very religious and very practical. Apart from the invention of a very simple calendar whose use spread across the whole of the ancient world, we find no contribution of the same importance as those we shall go on to mention. Neugebauer put this very clearly: 'Ancient science was the product of a small number of people; and it so happens that they were not Egyptians.'[12] For the Babylonians, contemporaries of Egyptian civilization and reasonably close, the answer is more complex. Until the mid twentieth century, it was believed that their contribution to astronomy was solely in terms of observations. We know now that their impressive accumulation of data, transcribed onto thousands of clay tablets, enabled them to display certain regular movements of the heavenly bodies.

In fact, starting from the second millennium BCE, the Babylonians began systematically to record the positions of the Moon and Sun in relation to the constellations. For the planets, they noted the angle between the sightlines Earth–planet and Earth–Sun. They observed that for Mercury and Venus this angle varied only little: between 0° and 24° for Mercury, and 0° and 44° for Venus, leading them to believe that these planets were in some way connected to the Sun (turning around it?). For Mars, Jupiter and Saturn, on the other hand, this angle could vary between 0° and 180° (opposition). Basing themselves on the regularities of these results, they managed to predict the moment and the region of the sky where these exceptional events would take place. The Babylonians were also able to predict the date and (with greater uncertainty) the zone of visibility on Earth of lunar eclipses, basing themselves on the results of some sixty eclipses recorded between 695 and 67 BCE. The legendary quality of the data collected by the Babylonians, however, can be disputed. Ptolemy (c. 90–180 CE) repeatedly complained of the mediocre quality of their observations of planetary positions.[13] This is all the more likely in that climatic conditions in Mesopotamia are highly unfavourable, on account of mists and desert sandstorms that particularly affect observations close to the horizon.

This idea of Babylonian astronomy as no more than observation has radically changed. As new tablets were successively deciphered, it was discovered that the Babylonians had also developed mathematical models of the solar system. This belated deciphering of the tablets explains why great historians of ancient astronomy, such as Delambre, were unaware of this foundational work of theoretical astronomy. We know today that although astronomy remained purely qualitative until the Assyrian period (early sixth century BCE), there was then a turn towards a mathematical description, and during the last three centuries BCE the Babylonians developed theories of the movement of the Moon and planets of as good quality as their Greek colleagues. Their methodology, however, was totally different. The Greeks studied the continuous movements of the planets on the basis of geometrical considerations, whereas the Babylonian scholars noted series of singular events (full Moon, eclipse, retrogradation) and searched these for arithmetic regularities.

Astronomy in Asia

China

Perhaps more than in other civilizations, astronomy/astrology played a major political role in China. At certain times there was even a Bureau of Astronomy under the Minister of Rites. The Chinese did not share Aristotle's assumptions as to the immutability and perfection of the supralunary world, and observed minutely and without fear the 'guest stars', that is, comets and supernovae, seeing these as messages from the gods that astrologers had to interpret. They left precise records of these that are useful for modern astronomers, as Western archives are rather vague on this subject. Similarly, they dared to observe and note spots on the Sun (unthinkable for Aristotle) fifteen centuries before Galileo. But it was not until the fourth century CE, six hundred years after Hipparchus, that the Chinese discovered the precession of the equinoxes,[14] which they called the 'annual difference' (between the solar year and the sidereal year). They made no attempt, as the Greeks did, to establish a general model. It was the Jesuits who introduced into China for the first time the diurnal motion of Earth and heliocentrism, in the seventeenth and eighteenth centuries. The Jesuits sought to use their higher quality of prediction to gain influence at the Chinese court, with varying success.[15] As against their chronology, Chinese astronomic works had practically no influence on Europe.

India

As we have said of Chinese astronomy, Indian scholars also showed little interest in developing general models. But we cannot overlook the work of the mathematical and astronomical genius Âryabhata (476–550), author of a treatise in verse, the Âryabhatîya. This contains, among other things, extraordinarily precise estimates of π and the radius of Earth, the resolution of algebraic equations, the formulation of trigonometric theorems (Âryabhata was the inventor of the sine), etc. He proposed a model of Earth's diurnal rotation. To illustrate the relativity of motion, he gave the famous argument of the apparent movement of the shore,[16] a thousand years before Galileo: we know that it is the boat and not the land that changes position. The translation of the verse from the Sanskrit reads, 'Just as from a moving boat one sees the mountain shift in the opposite direction, so the stars that are just as immobile go directly west under the equator.'

Âryabhata is even sometimes attributed a conception of heliocentrism, but there is no source that confirms this.[17] It was not until the Kerala school of Indian astronomy, between the fourteenth and sixteenth centuries, that the mathematician Nilakantha Somasutvan (1444–1550) proposed a partially heliocentric model.[18] There is no reason to believe that Tycho Brahe, who developed a very similar system a century later, was familiar with his work.

The Astronomy of Ancient Greece

The first advances of astronomy in ancient Greece are generally dated around the beginning of the sixth century BCE. This was when the first models were proposed that attempted to explain the movements of the stars. Unfortunately, the overwhelming majority of original texts disappeared or are known only secondhand.[19] For this reason, the origins of several discoveries are debatable or uncertain, and the same even holds for almost all dates of birth and death of the scholars in question. The great revolution undertaken by Greek astronomers was the desire to understand the origin of the phenomena they observed, over and above the possibility of predicting them. This was not yet the emergence of science in the present-day sense, inasmuch as the aesthetics of an argument were more important than agreement with experimental evidence, and the tool of mathematics still very rudimentary, essentially just limited to geometry.

Thales of Miletus (sixth century BCE), one of the 'seven sages' of ancient Greece, is often presented as the first known philosopher to be interested in astronomy. Diogenes Laertius, in the early third century CE, from whom we have information on a number of Greek philosophers (though often very superficial and even rather dubious), writes that Thales was 'the first [who] traced the course of the Sun from one solstice to the other, and showed that the Moon was a 120th part compared with the Sun'.

We do not know how Thales made this estimate. Again according to Diogenes Laertius, Thales had a view of the cosmos that was still very archaic, probably taken from the Egyptians: the Cosmos was a sphere of water enclosing a bubble of air, the Earth, that floated at the bottom of this sphere.[20] It was to Anaximander, who was probably a disciple of Thales, that the revolutionary idea is attributed of a compact (cylindrical) Earth surrounded on all sides by a sky containing the stars, the Moon and the Sun (in that order). The notion of absolute up and down, and thus of pillars or animals on which our planet rested, was abandoned. The cosmology of the Miletus school eschewed any divine intervention, which is why Carlo Rovelli, for example, sees it as the origin of scientific thought.[21] The school of Pythagoras (c. 570–480 BCE) continued this tradition. Among the most striking results of the Pythagoreans were the decomposition of the Sun's movement into a diurnal rotation from east to west, and an annual motion in the opposite direction.[22] We should also mention the hypothesis of Earth's round shape, despite this being based solely on considerations of symmetry. Proofs only came later: Aristotle, in particular, demonstrated this by the crescent shape of the shadow that Earth casts on the Moon during an eclipse, which only a spherical Earth could produce. Taking certain images from the Pythagoreans, Plato (427–348 BCE) constructed a total image of the cosmos. He imagined an infinite space in which a finite universe was immersed, made up of homocentric (i.e., with the same centre) spheres. The innermost and immobile sphere was Earth. The outermost one carried the fixed stars, turning on itself from east to west, which explained their diurnal motion. The other spheres were occupied, from outer to inner, by Saturn, Jupiter, Mars, Mercury, Venus, the Sun and the Moon. They turned around an axis perpendicular to the ecliptic, but at different speeds.

Plato's model offered an overall view of the universe. But though aesthetically satisfying, it was incapable of explaining a good many observations, in particular the retrograde movements of the planets (see below). As we shall see, this led one of his disciples, Eudoxus of Cnidus (c. 408–355 BCE), and then Aristotle, to increase the number of homocentric spheres, up to fifty-five in Aristotle's case!

The Aristotelian carapace

The astronomers of antiquity, from Plato on, had a dual vision of the world. There was the 'perfect' world of the stars, eternal and infinitely regular: the stars all described rotary movements around the Pole star, observable on a timescale of a few hours; the constellations followed an annual course and returned to the same positions in the sky; the Sun also described a circular movement in relation to the fixed stars. This supralunary world was governed by a rationality that it was the mission of mathematics to express. At the base of the hierarchy there was the world in which we live, changing and corruptible.[23] With Aristotle, these notions acquired a theoretical status, the value of a principle. From now on, movements that were irregular, violent, unbalanced, were characteristic of the sublunary world. Here, a body cannot persist in movement without a force that sustains this; its speed is proportional to this force.[24] The supralunary world, for its part, is animated by natural, circular movements, without beginning or end since they are enclosed on themselves, images of a perfect motion that there is no need to question. We find a good summary of this from the scholar Abu'l-Barakat al-Bagdadi (c. 1070–1152), known as 'The unique of his time', even though he was critical of Aristotle[25]:

> The circular movement is also particularized by the fact that it is perfect and admits no increase, it is identical to itself, it neither strengthens nor weakens, compared with natural movement [i.e., free fall] which increases at the end, or violent movement [the firing of a projectile] which is stronger at the beginning.[26]

The planets do indeed present these strange retrogradations, but, since they belong to the supralunary world, they have to be accounted for by a combination of uniform circular motions.[27] In modern terms, we would say that Aristotle drew from this the idea that the two worlds must be described by two physics. The low (sublunary) world is composed of the four elements: earth, water, air and fire. The two former, the 'weighty' elements, occupy at rest their natural place, that is, the lowest possible, or tend to reach this by following a straight line. Fire likewise follows a straight line, but a rising one. As it is the nature of bodies that explains the nature of their movement,[28] the celestial bodies do not fall, since their composition, the quintessence, is different. In the supralunary world, therefore, rather than gravitation there is the constraint of *using only circular motions*, or a combination of circular motions, which alone are perfect on account of their symmetry. This constraint soon acquired the status of a theoretical tool. It was accepted for two thousand years, not being abolished either by Copernicus or Galileo. The ancient astronomers were as attached to this constraint as modern scientists are to the conservation of energy or the maintenance of certain symmetries. It was Kepler who first challenged it. We shall come back to this episode, which is of major importance for the evolution of ideas that led to Newton's mechanics.

Let us venture a parallel that is clearly superficial: we have experienced the stifling weight of 'Stalin's thought', followed by that of 'Mao Zedong thought', but what are thirty years compared with fifteen centuries![29] Aristotle's thought,[30] which dominated and even crushed every field of intellectual life, had contrasting effects: it systematized the necessity of logic in reasoning, but it also introduced a form of circularity still current enough for Molière to make fun of in the seventeenth century, when he had Bachelierus say (in Latin) in *Le Malade imaginaire* that 'opium causes sleep because it has a dormitive virtue whose nature is to dull consciousness'. Kepler was still in the Aristotelian tradition when he explained the movement of the planets by a motive spirit (which he changed to motive force). And in the twentieth century, what of Bergson, who introduced a vital force (or impulse) to account for life?

The question to be resolved then could be summed up as follows: how to explain the irregular movement of the planets, the Sun and the Moon, on the basis of circular motions with a constant angular velocity, and with Earth at the centre of the system?

Eudoxus' model

Until the end of the fourth century BCE, Eudoxus' theory of homocentric spheres was the dominant astronomic model. Earth stood immobile at the centre of a system of twenty-seven spheres, all centred on it, which carried the Moon, the Sun, the five planets that were known at this time (the wandering stars) and the fixed stars. Without embarking on a complex description of this model, we can just say that, to describe the movement of each of the first two bodies, Eudoxus introduced three spheres, then a further four for each of the five planets. Two spheres accounted for their average motion and two others for the retrogradations. Finally, the outermost sphere simulated the diurnal rotation of the stars, turning from east to west in twenty-four hours. The system was ingenious and complex: each of the spheres had its particular direction and axis of rotation. The combined movement of all these spheres generated for each of the planets a figure-of-eight curve known as the *hippopede*. For the first time there was an explanation of the retrogradations.

As distinct from the rest of his contemporaries, Eudoxus was not content with formulating a theory, but proposed numerical values making it possible to calculate the expected positions of the heavenly bodies at precise moments. We know from Simplicius, who lived in the sixth century CE, the values that Eudoxus used for the rotation periods of the second sphere containing the five planets, periods that differed very little from the present values for the rotation periods of these planets around the Sun.[31]

Eudoxus' model made it possible to account quite faithfully for the average positions of the wandering stars, provided that appropriate values were used for the various parameters (angles and speeds of rotation). It did however present certain defects inherent to the model's very foundations, including the following:

1. The variations in the apparent diameter of the Moon are inexplicable, as are also the variations in the brightness of the planets, phenomena that were correctly interpreted later as due to variations in the distances of these bodies from Earth.
2. The uneven duration of the seasons was likewise unexplained.
3. The *hippopedes* only very roughly imitated the actual trajectories of the retrogradations.

The two first points are a direct consequence of the hypothesis of spheres with the same centre. These difficulties played a large part in this theory being abandoned in favour of other geometrical constructions that allowed a greater degree of freedom.

For Eudoxus, these spheres clearly remained simply a mathematical tool that 'saved the appearances', that is, described the apparent movement of the stars and planets on the basis of a combination of circular movements alone. In no case did these spheres have a real existence. This was however Aristotle's position. Still much later, in the Middle Ages, the idea that the celestial spheres were physical objects was bitterly defended, which gave rise to several problems (including theological ones: how could the angels cross them, and how was the Ascension of Christ possible across these spheres?).

The insurmountable difficulties of this model gradually led to its displacement in favour of other geometrical variants. In any case, the basis was Aristotle: there are only circles, even if their centres no longer coincide.

Before discussing in more detail the genesis of Ptolemy's model, the most sophisticated of these, which remained current for more than 1,500 years, we should mention one striking exception, but which did not have any sequel: the heliocentric model of Aristarchus.

The Aristarchus parenthesis

Aristarchus of Samos (310–230 BCE) is generally seen as having been the first to propose a heliocentric system of the cosmos. Curiously, however, he did not draw the consequence of this for the irregularities in the movement of the planets. His work was lost, and we know his ideas only by way of a brief commentary by Archimedes (c. 287–12 BCE), who says in *The Sand Reckoner* (or *Psammites*):[32]

> '[U]niverse' is the name given by most astronomers to the sphere whose centre is the centre of the earth and whose radius is equal to the straight line between the centre of the sun and the centre of the earth … But Aristarchus of Samos brought out a book consisting of some hypotheses … that the fixed stars and the sun remain unmoved, that the earth revolves about the sun in the circumference of a circle.[33]

Aristarchus may have fallen victim to persecution (in this respect, too, presaging Galileo) for having dared to place the Sun at the centre of the universe, with a consequent risk of troubling the peace of the gods.[34]

The model of Apollonius of Perga

Beyond its historic interest, the model of Apollonius (262–190 BCE) served as a theoretical foundation for that of Ptolemy three centuries later. It constituted, in a way, a version of this for beginners, relatively easy to understand. There followed a century later the first improvement, due to Hipparchus, at the price of a certain complexity. The ground was now prepared for Ptolemy's improvements, but this time at the price of quite enormous complexity, which cannot be fully explored in detail in the limits of the present book.

As a preliminary remark, in all the models that we shall explain here through to that of Copernicus, the movement of the heavens is always described in abstraction from what in modern terms we know as the diurnal rotation of the Earth, which was then seen as the rotation of the sphere of fixed stars. It involved in fact the projection onto this sphere of a movement of uniform rotation.

In Apollonius' model, the Sun moves in a circle centred on a fixed Earth E, while each of the five planets P describes a small circle, known as an epicycle, whose centre C traces a larger circle called the deferent (Figure 18). This deferent represents, as it were, the average trajectory of these bodies, around which the actual trajectories undulate. All these circles move in the plane of the ecliptic.

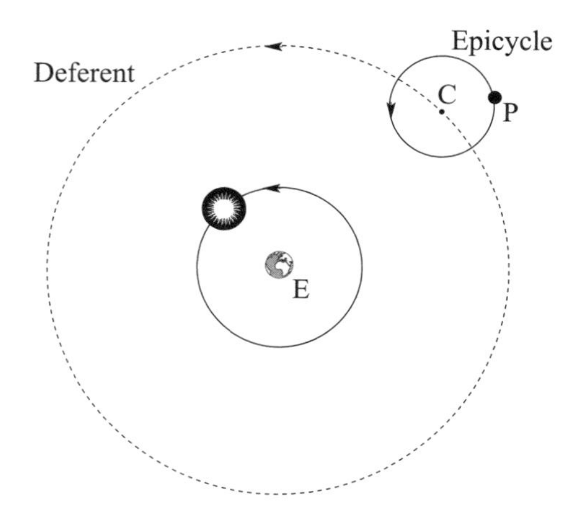

Fig. 18. The system of epicycle and deferent in the geocentric model of Apollonius. Here the Sun describes, without an epicycle, the first circle with Earth as its centre. Planet P describes an epicycle whose centre C itself turns on its deferent with Earth as its centre.

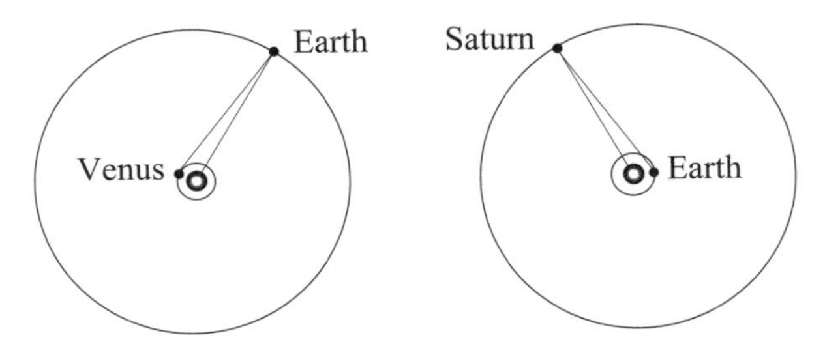

Fig. 19. Trajectories of an inner planet (Venus, on the left) and an outer one (Saturn, on the right). The distance Earth–Venus is always close to the distance Earth–Sun, just as the distance Earth–Saturn is always close to the distance Sun–Saturn. (The diagram is not to scale.)

Where could such an idea have come from?

The first attempt at a theory of epicycles was undoubtedly due to Heraclides of Pontus (c. 388–315 BCE), who investigated, like the Babylonian astronomers before him, the movements of Mercury and Venus, the two planets closest to the Sun. As seen from Earth, these planets are never very distant from the Sun, oscillating from one side of it to the other (Figure 19). Although none of his writings have been preserved, Heraclides was certainly the first to suggest that Mercury and Venus actually turned around the Sun. Their movement was thus made up of a principal motion along the deferent circle, in which they followed the Sun around Earth, and a correction due to the rotation of the planet around this deferent in an epicycle. As we shall see, this can account for the retrogradations of these planets against the background sky.

But retrogradations are observed also for the three other planets, despite their seeming not to be bound to the Sun, given that in each case the angle Planet, Sun, Earth can be up to 180 degrees. Hence in all probability Apollonius' idea of representing the annual movements of Mars, Jupiter and Saturn also as a composite of two motions: a principal motion of large amplitude around Earth (a deferent), and a motion in a small circle (epicycle) centred on the deferent.[35] This time, however, the centre of the epicycle is not occupied by any real object, as against the cases of Mercury and Venus, where it is occupied by the Sun.

The heliocentric explanation of Apollonius' model

We have shown how Apollonius had been led to develop his model with the knowledge of his time; we shall now explain its construction, and show how the modern (heliocentric) point of view enables its success to be understood. Figure 20 illustrates how it works. For the planets that we now know to be inner ones, the construction is clear: the epicycle is centred on the Sun, whose trajectory is the deferent of the planet. Let us now consider an outer planet such as Jupiter, marked P on the right of Figure 20. Taking Earth E as origin, the Sun S describes a circle, and, as we now know, Jupiter (perceptibly) describes a circle centred on the Sun. We therefore construct the segment ES (of fixed length), then the segment SP (likewise of fixed length). But we can proceed in another way: constructing a segment ES′ parallel and equal to SP, then a segment S′P parallel and equal to ES. Since SP is of constant length, the same holds for ES′, therefore the point S′ describes a circle centred on Earth, the deferent whose radius is equal to the radius of the planet's orbit around the Sun. In the same way, S′P is of constant length: P moves in a circle centred on S′, an epicycle whose radius is equal to the Earth–Sun distance.

The epicycle thus obtained represents, as it were, the transfer of the Sun's motion (seen from Earth) to the neighbourhood of planet P.

We also note the following strange aspect. The first construction only involves 'real' objects: the motionless Earth, the Sun, the planet and their trajectories.[36] The second introduces a fictitious and abstract point, the centre of the epicycle, moving on a fictitious trajectory, the deferent centred on the Earth.

To simplify the explanation, we have taken the radiuses of the deferents of the inner planets and of the epicycles of the outer planets as the average distance from the Sun to Earth. We have also taken the radiuses

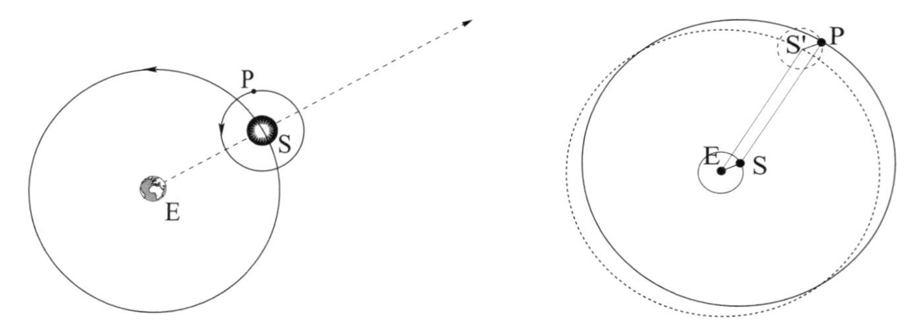

Fig. 20. On the left, the construction of the epicycle for an inner planet in Apollonius' model. The axis ES has to pass through the Sun, making one rotation in a year. The vector radius joining the centre of the epicycle and the planet is always parallel to the Earth–Sun radius. (Not to scale.)

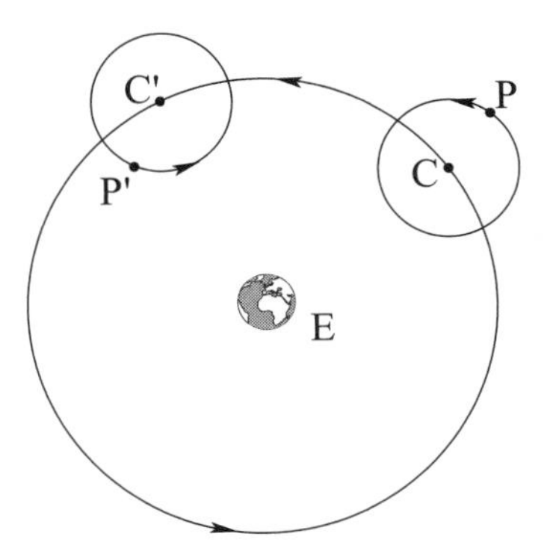

Fig. 21. Positions and velocities of a planet at two dates in the deferent-epicycle system.

of the deferents of the outer planets as equal to their average distance from the Sun. But the same pattern could be obtained by multiplying all these lengths by the same factor. In actual fact, Apollonius did not know the absolute distances between these bodies. His model could only give angles and ratios of distances.

This schema can describe quite naturally the retrogradations against the background sky. Figure 21 shows the positions P and P′ of a planet at times *t* and *t*′, corresponding to two positions of the centre of their epicycle, C and C′. It is apparent that at time *t* the velocity, as seen from Earth, results from the addition of the speeds of C and of the movement around the epicycle, while at time *t* these two speeds counter each other. At P, the angular velocity will appear greater than average, while at P′ it will appear smaller, even to the point of retrogradation.

Hipparchus of Nicea's improvement

With Hipparchus of Nicea (c. 190–120 BCE) we reach the real beginning of quantitative astronomy, especially as far as the positions of the fixed stars and the movements of the Sun and Moon are concerned. His name is less well known than that of Ptolemy, thanks to whom we know a good part of his work, but his contribution to astronomy was certainly more fundamental.

The author of a catalogue that gave the positions of more than eight hundred stars with a high degree of precision, Hipparchus was first of all a remarkable observer. His measurements were taken with an instrument called a *dioptra*, a kind of theodolite that he helped to perfect. The *dioptra* consisted of a long sight equipped with two finders: the one facing the observer had a fixed pinhead hole, while the other, towards the object aimed at, was pierced with two holes and could move along the sight. It was possible in this way to measure the angular distance between two stars.

According to Ptolemy, Hipparchus elaborated his catalogue with the object of enabling future generations to note slow movements on the part of stars or even the appearance of new ones, so that he presumably no longer believed in the invariability of the fixed stars. His observations on the positions of stars also led him to the discovery and quantification of the phenomenon of the precession of the equinoxes (see p. 61 above), one of his major contributions. Not to forget that we likewise owe him the introduction of trigonometric functions.

The defects of Apollonius' model that Hipparchus revealed were:

- It predicted seasons of equal length on account of the uniform circular movement of the Sun.
- It predicted strictly periodic retrogradations.

The solution that Hipparchus found was to slightly decentre the Earth.[37] The Sun continued to turn at a uniform speed, but in a circle whose centre O no longer coincided with that of Earth (see Figure 22). So the angular speed of the Sun as seen from Earth (the θ variation) was no longer uniform, being a maximum when the Sun was closest to Earth (perihelion) and a minimum when it was farthest (aphelion).

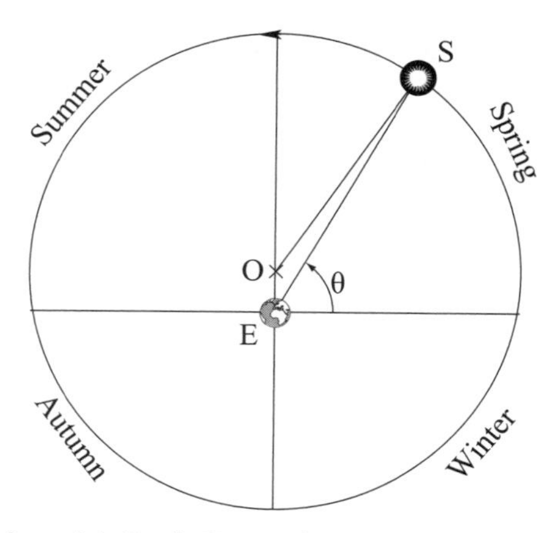

Fig. 22. Hipparchus' model. Earth E is no longer at the centre O of the Sun's deferent. As the Sun moves at a constant speed, the deferent now divides into four seasons of unequal duration (see text).

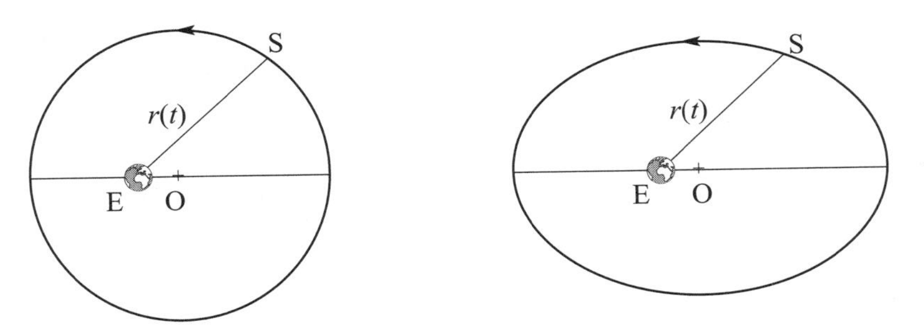

Fig. 23. On the right, how the movement of the Sun S appears in relation to Earth E. The eccentricity of the ellipse is exaggerated here for illustrative purposes. On the left, its simulation in Hipparchus' model, obtained by a decentring of the Sun (likewise exaggerated).

The modern explanation

We know today that Earth actually moves around the Sun in an ellipse, which also means that, taking Earth as origin, the Sun describes an ellipse as shown on the right of Figure 23. The Sun–Earth distance is accordingly not constant. One way of simulating the variation of the radius $r(t)$ joining Earth to the Sun in the course of its rotation is to decentre the Sun, as indicated on the left side of Figure 23.

Ptolemy's Model

The great astronomer of the late Hellenistic period was Claudius Ptolemy of Alexandria, who was also its great geographer. The dates of his birth and death are uncertain (c. 90–c. 170 CE), but his astronomic observations can be dated precisely, between the years 127 and 141, in an Alexandria that was already under Roman occupation. His *Syntaxis mathematica*, or *Almagest*, a mythical book of ancient astronomy, served for more than fourteen hundred years as an absolute reference for celestial movements. Ptolemy often cites his illustrious predecessor Hipparchus, so that very often his own contribution is unclear. The *Almagest*, written around the year 150, has survived thanks to its translation into Arabic. It was also thanks to Arabic astronomers that this text became known in the West.

Ptolemy's great contribution to astronomy was to complete the elaboration of the geocentric mathematical model of the movement of the Sun, Moon and planets. For the first time, it was possible to account for past observations and predict future ones, with a precision that matched the quality of data of the time. This model was more sophisticated than that of Apollonius in two ways:

1. It generalized Hipparchus' decentring to all the deferents.
2. It took into consideration the fact that the plane of the planetary orbits (and that of the Moon) is not strictly that of the ecliptic.

1. Not only did Ptolemy decentre *all* the deferents, but he no longer assumed that the movement of centre C of the epicycle along the deferent was uniform; he took what he called an 'equant' point E, symmetrical to T in relation to O, the centre of the deferent (see Figure 24). Thanks to this 'brilliant cheating',[38] which he did not justify, he managed to make the angle θ of the radius EC vary uniformly. Ptolemy thus achieved the tour de force[39] of keeping only circular and uniform motion, but rather more of one than the other …

2. It remains that the planes of the deferent and the epicycles do not coincide exactly with that of the ecliptic. Accordingly, Ptolemy introduced for each of the planets, with the help of little circles perpendicular to the ecliptic, ad hoc orientations of the planes of the deferents that made the model coincide better with observations. The insurmountable difficulty he was confronted with is that, in his model, the plane of the deferents cuts that of the ecliptic along a line (the nodal line) passing through Earth, whereas it should pass through the Sun. Copernicus could not improve on this, and it was only Kepler who resolved the problem.

Throughout the Middle Ages, Ptolemy's model was thus treated as the absolute point of reference. His astronomical observations,

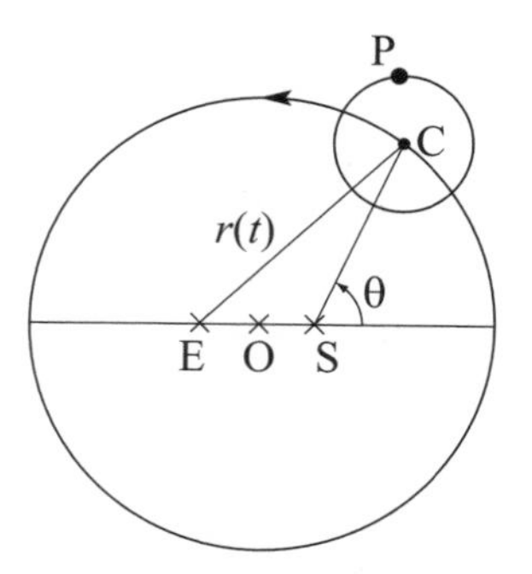

Fig. 24. Decentring of the deferent of centre O in Ptolemy's model. The angle θ varies linearly over time.

however, had a different fate. As early as 1008, Ibn Yunus (c. 949–1008), the Cairo author of astronomical tables (see the section below on Arab astronomy), noted that several of Ptolemy's observations contained mistakes. Much later, in modern times, Delambre (1749–1822)[40] wondered whether Ptolemy actually did make observations, and in his *Histoire de l'astronomie du moyen age* accused him of having fabricated some of his solar observations.[41] In recent times, Robert Russell Newton, a physicist and acclaimed specialist in the secular accelerations of the Moon and Earth, published a book, *The Crime of Claudius Ptolemy*, in which he maintained that Ptolemy manipulated not only the observations he claimed to have made himself, but also those of other older astronomers, including some from Babylonian sources.[42] This book provoked a vigorous argument among historians of science, with those possessing a solid grounding in mathematics and physics tending to support Newton's claim. Whatever the truth, everyone agreed that Ptolemy at least 'selected observations that best matched his theory',[43] a practice that perhaps was not seen at the time as actually fraudulent.

Criticism in the Arabic-Muslim World

As with Babylonian astronomy, it was only the work of Neugebauer in the mid twentieth century that showed how the role of Arabic astronomy[44] cannot be reduced to that of rescuing and transmitting Ptolemy's legacy; Western specialists had previously been familiar only with those Arabic texts translated into Latin.[45] It must be remembered that the European Middle Ages corresponded to a golden age in the Islamic world, a period that saw not only the flourishing of astronomy, but also that of mathematics, geography and medicine.[46]

Arabic astronomers, for utilitarian reasons, were more interested than their Greek predecessors in the ability of models to predict celestial movements: knowledge of the positions of the Moon was necessary for determining the start of Ramadan, and the requirements of long-distance navigation and positioning in the desert demanded precision. Well acquainted with Indian writings, they notably perfected trigonometry (including spherical) and geometry; Al-Khawârizmi[47] in the eighth century popularized what we call Arabic numerals, which in fact came from India, and developed algebra. In terms of observation, these astronomers introduced, four hundred years before Tycho Brahe, a study of the sky followed up over very long periods; they perfected the Greek astrolabe and honed instruments of amazing precision.

Among the continuers and subsequent critics of Ptolemy in the Arabic world, names that stand out are those of Jabir ibn Aflah (1100–60) in Spain, Nasir al-Din al-Tusi (1201–74) in Persia and Ibn al-Shâtir (1305–72) in Syria. The second of these is remembered for his model of the *Tusi-couple*, an ingenious mechanism that yields a straight-line movement on the basis of circular movements alone.[48] The theoretical importance of this is considerable, as it could now be assumed that the equant moved in a straight line without derogating from Aristotle's sacrosanct principle of the necessary circularity of supralunary movements. The remarkable precision that the new instruments allowed on the positions of the stars (in the order of a minute of arc) led these scientists to growing critiques, not only of Ptolemy's results, but also of his model. For example, in a celebrated text entitled *Doubts on Ptolemy*, Ibn al-Haytham (965–1039) criticized the master's model for certain bad predictions (for example, that the apparent diameter of the Moon would vary by a factor of 4, which is clearly fantastic), and above all its conceptual incoherence: how could one at the same time claim fidelity to Aristotle's principles and violate them so blatantly with eccentrics and equants? A change of model was therefore needed. Some writers even proposed returning to a modified version of Eudoxus' model, but without success. The culmination of these attempts was the multi-epicycle model of Ibn al-Shâtir, which gave perfect agreement[49] with all the then known data on the movement of the planets.

Penetrating as they may have been, it would be wrong to believe that these criticisms led to the geocentric model being challenged. But as recent work has shown,[50] they were known to Copernicus and were essential to his development. It remains an open question why the Copernican revolution was not brought about by the scientists in the Arabic world who were so well prepared for it.[51] As early as the eleventh century, the great Persian scientist Al-Biruni (973–1048), who was familiar with the works of Âryabhata, wrote:

> I have seen the astrolabe invented by Abu Saïd Sijz. I appreciated it greatly, as it is conceived on the idea some people have that the apparent movement comes from Earth's own motion and not from that of the heavens. On my life, that is a problem I find hard to resolve ... Whether it is Earth or the heavens that are

> in movement changes nothing. In neither case does this affect astronomy; it is up to the physicist to envisage a possible refutation.[52]

Finally, Al-Biruni followed the argument of the Indian astronomer Varâhamihira: 'If such were the case [that Earth moved], a bird would not return to its nest, as it would be swept to the west.'

Tycho Brahe's Model

The Dane Tycho Brahe (1546–1601) was certainly the last great observer of the stars with the naked eye, and probably one of the most meticulous. He had an illustrious disciple, Kepler. His model appeared after the heliocentric model of Copernicus. Appendix D shows that it is kinematically equivalent. But since it is generally presented as the last geocentric model, it is logical to discuss it now.

All the planets except Earth, which is fixed, turn around the Sun. The Sun, for its part, turns around Earth (Figure 25). Though formally geocentric, this could also be called a physically heliocentric model, but *as seen from Earth*. Rather than geocentric, it is *Earth-centred*. The whole of the solar system is then described far more simply than with

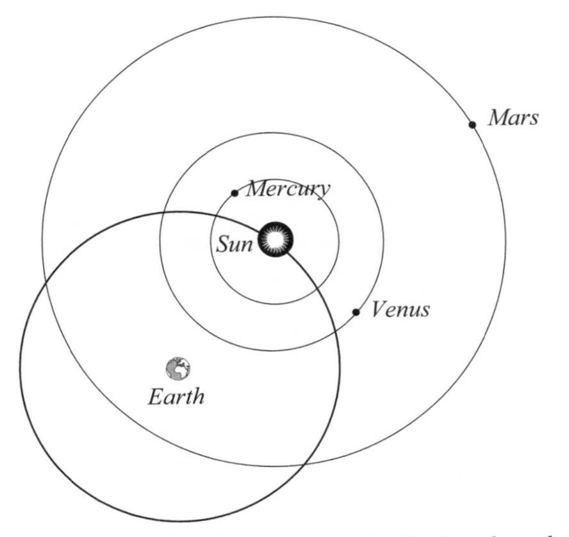

Fig. 25. Tycho Brahe's model. Earth is at the centre. The Sun describes a circle with a radius taken as 1. Mercury, Venus and Mars each move in a circle centred on the Sun, with respective radiuses b = 0.4, 0.7 and 1.5.

Ptolemy, all the planets having the same deferent (the trajectory of the Sun). Tycho Brahe's system accounts for the phases of Venus, later discovered by Galileo (in 1610), which Ptolemy's model cannot explain. It also keeps the advantage, along with all geocentric models, of taking as its starting point observations made from Earth.

Tycho Brahe had his famous Uraniborg observatory built close to Copenhagen, equipped with the best instruments of his day. Despite being conducted with the naked eye, his measurements enabled him to establish a remarkable catalogue of the angular positions of celestial bodies. Kepler understood the importance of having available data of this quality, and in fact all his own theoretical work rested on this. He had to be well convinced of the exactness of his results to maintain, for example, that the trajectory of Mars is an ellipse and not a circle, given the very slight eccentricity of this planet (see below, Figure 29).

The year 1572 saw the appearance in the sky of a new and brilliant object (now known as Tycho's Nova), followed by the great comet of 1577. The great merit of Tycho Brahe was to have understood that both of these bodies belonged to the supralunary domain (they were sufficiently remote to have no observable diurnal parallax, as opposed to the Moon and the planets). Aristotle's principles were then doubly infringed: how could an object (the nova) be born, then disappear, in a heaven supposedly immutable, and how could the visibly non-circular trajectory of the comet be accepted? In any case, the planets could not rest on crystal spheres, as the comet would have broken these in crossing them. Tycho Brahe thus signalled the end of the 'corporeal spheres'.

Tycho Brahe's contribution was not so much his model[53] as:

1. The quality of his observations, which enabled Kepler to discover his laws;
2. His discovery of 'impure' supralunary objects, which marked the start of the repudiation of Aristotelian physics.

CHAPTER 5

The Construction of Heliocentrism

> Philosopher: Your Highness, my esteemed colleague and myself are supported by no less an authority than the divine Aristotle.
>
> – Bertolt Brecht, *The Life of Galileo*

The image of 'dwarfs on the shoulders of giants' illustrates the filiation running from Copernicus to Kepler, then from Kepler to Newton. We shall not be discussing Galileo here: this other giant, however indispensable for Newton's future development, as well as in the defence, illustration and spread of heliocentrism, did not modify Copernicus' model in the way that Kepler had done.

Two ruptures stand out in the history of ideas: firstly, the break with Aristotle's thought, then the break with a literal reading of the Bible. This distinction, despite being artificial (anachronistic), does illuminate the course of thought.

Nicolas Copernicus (1473–1543), the Polish astronomer, is universally known for having had the boldness to expel Earth from the centre of the universe. He proposed a heliocentric model whose main advantage at that time was its elegance: the retrograde movement of the planets followed naturally from it, without the artificial construction of epicycles, deferents and other equants. Unfortunately, in its simple form (Figure 27), Copernicus' model had less predictive power than that of Ptolemy. Besides, wishing to preserve the Platonic principle of circular orbits, Copernicus was forced to reintroduce, however marginally in his case, a system of epicycles, the only function of which was to better reproduce the actual trajectories (which in fact were ellipses, as we know thanks to Kepler) with combinations of circles. His book *De revolutionibus orbium coelestium*,[1] completed in 1530, would only be published a few days before the death of its author (a prudent move). Here is his own most concise summary of his future theory, published in

his *Commentariolus* (probably dated 1513), which remained fairly confidential:

> **All the apparent movements that are observed in the firmament are due to the movements of the Earth and not of the firmament.**

This says the essential.

Copernicus' model, by abandoning the principle of geocentrism, makes it possible to simplify the description of the solar system enormously.

Nearly a century later, Kepler's (1571–1630) break with the Aristotelian dogma of circular trajectories led to the relegation of Copernicus' complex system of residual epicycles, while also gaining in precision: the planets (including Earth) now travelled in ellipses (though almost circular ones), with the Sun occupying in each case one of the two foci. Kepler put forward three laws of motion, universally known today as Kepler's laws (see Glossary).

Half a century after Kepler, Newton (1624–1727) showed, in the context of a general reflection on the causes of motion, that these three laws resulted from just one: the law of universal attraction (combined with the general law of motion).

Backed by the mathematical analysis that Newton had powerfully developed, modern physics was born.

Copernicus

Copernicus' model is described by seven postulates, contained in his *De revolutionibus orbium coelestium*:

First postulate: There is no one centre for all the celestial orbs or spheres.

In other words, Earth is only a secondary centre, that of the orbit of the Moon. The second postulate explains the first:

Second postulate: The centre of the Earth is the centre, not of the universe, but only of gravity and of the lunar sphere.

This is still an Aristotelian view of gravity, despite the fact that Copernicus' own model makes this untenable: what is 'up' and what is 'down'?

Third postulate: All the spheres encircle the Sun, which is as it were in the middle of them all, so that the centre of the universe is near the Sun.

In actual fact, since Earth's trajectory is elliptical, Copernicus could not account for it by a circle exactly centred on the Sun. He had to introduce a neighbouring fictional point as centre, the 'average Sun'.

Fourth postulate: The ratio of the Earth's distance from the Sun to the height of the firmament is so much smaller than the ratio of the Earth's radius to its distance from the Sun that the distance between the Earth and the Sun is imperceptible in comparison with the loftiness of the firmament.

This fourth postulate addresses the recurrent question of believers in an immobile Earth: if Earth is in annual movement around the Sun, in the motionless heavens, one should see the fixed stars (with a correction made for diurnal motion) at different angles when Earth occupies its most extreme positions, that is, at six-monthly intervals. In scientific terms, this is the annual parallax of the stars. Why has this parallax always been measured as zero? The answer is given in this postulate: the base of the diameter of the terrestrial orbit is extraordinarily small in relation to the distance of the fixed stars from the Sun,[2] making the parallax too small to be measured.

If Copernicus' model does not yield absolute distances from the planets to the Sun,[3] it does make it possible to know with great precision their ratios to the Earth–Sun distance, a distance today called the 'astronomical unit', or AU. There was already here a prescience of those 'boundless spaces' that frightened Pascal so.

Fifth postulate: Whatever motion appears in the firmament is due, not to it, but to the Earth. Accordingly, the Earth together with the circumjacent elements performs a complete rotation on its fixed poles in a daily motion, while the firmament and the highest heaven abide unchanged.

This explains the succession of day and night.

Sixth postulate: What appear to us as motions of the Sun are due, not to its motion, but to the motion of the Earth and our sphere, with which we revolve about the Sun as [we would with] any other planet. The Earth has, then, more than one motion.

This explains the apparent annual motion of the Sun.

Seventh postulate: What appears in the planets as [the alternation of] retrograde and direct motion is due, not to their motion, but to the Earth's. The motion of the Earth alone, therefore, suffices [to explain] so many apparent irregularities in the heaven.

This is the model's great strength. The seventh postulate indicates that the retrogradations are the direct result of the description of a heliocentric universe when this is seen from Earth, provided that the speeds of the planets are greater the closer they are to the Sun (Figure 15). This already shows the appearance of the idea that radius of orbit and speed are related:[4]

> The size of the spheres is measured by the amount of time

Figure 27 sums up the situation.

Apart from the theological questions that we shall examine in detail below, there still remained major objections of physics: why do we not feel the movement of Earth's rotation on itself? Why do guns have the same range firing east or west? Here is Copernicus' response[5]:

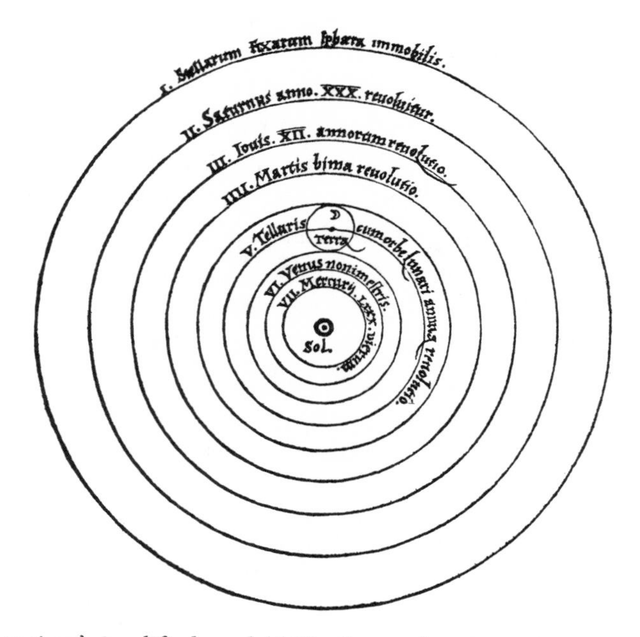

Fig. 27. Copernicus' simplified model.[6] The Sun at the centre is surrounded by six planets, including Earth, itself the centre of the Moon's motion. The outermost sphere is that of the fixed stars.

> But what is in accordance with nature produces effects contrary to those resulting from violence, since things to which force or violence is applied must disintegrate and cannot long endure. On the other hand, that which is brought into existence by nature is well-ordered and preserved in its best state. Ptolemy has no cause, then, to fear that the earth and everything earthly will be disrupted by a rotation created through nature's handiwork, which is quite different from what art of human intelligence can accomplish.

In other words, the effect of a 'natural' motion such as that of Earth's rotation should not be confused with a 'violent', artificial motion (away from equilibrium, we would say today). As Aristotle conceived things, there were two different physics. Copernicus proceeded with a different line of argument, almost *per absurdum*: if you believe that centrifugal force (in modern terms) acts on Earth, then all the more so would it act on the sphere of the fixed stars, whose radius is infinitely greater, and this would lead to a dispersion of the stars:

> But why does he not feel this apprehension even more for the universe, whose motion must be the swifter, the bigger the heavens are than the earth? Or have the heavens become immense because the indescribable violence of their motion drives them away from the centre? Would they also fall apart if they came to a halt? Were this reasoning sound, surely the size of the heavens would likewise grow to infinity. For the higher they are driven by the power of their motion, the faster that motion will be, since the circumference of which it must make the circuit in the period of twenty-four hours is constantly expanding; and, in turn, the velocity of the motion mounts, the vastness of the heavens is enlarged. In this way the speed will increase the size, and the size the speed, to infinity. Yet according to the familiar axiom of physics that the infinite cannot be traversed or moved in any way, the heavens will therefore necessarily remain stationary.

The text is less than clear, mixing an argument of physics with a more philosophical consideration. Playing on a paradox – the motion of the heavens, if there were such a thing, would be abolished by its own logic – this is an unconvincing and vaguely dishonest argument, when we know that for Aristotle the stars were not composed of *graves* (weighty material) and so not subject to what we today call centrifugal force. We should note that Copernicus gives a neat twist to his ultra-Aristotelianism here, equating the composition of all planets with that of Earth. In other passages he argues differently, speaking of the air being pulled by the motion of Earth, which in turns pulls the bodies in free fall.[7] In fact, Copernicus could not offer a precise response, as this would have required a knowledge of dynamics that only Galileo, and especially Newton, would provide later on.

Copernicus' simple model is incomparably more elegant than that of Ptolemy, developed with its ad hoc panoply of multi-equants, epicycles and other deferents; but unfortunately it is less precise. The need for these complications was explained by Kepler a century later: the trajectories are not circles, but ellipses. Copernicus, however, as a theorist, held so stubbornly to circles that he found the trick of Ptolemy's equants in contradiction with Aristotle, who authorized only uniform circular movements for the supralunary world. He thus had the idea of reintroducing epicycles into his model, though without their central function of explaining retrogradations, as Ptolemy had proposed. He simply gave them responsibility for small corrections: what turn in circular orbits around the Sun, in his system, are not exactly the planets, but more precisely, as indicated in Figure 27, little epicycles that carry the planets. And so (Figure 26) he simulated the ellipse by a combination of two circles uniformly travelled. Here the epicycles that were essential to Ptolemy bring only a small improvement, making his model almost as predictive as that of his illustrious predecessor. We just note that only Earth is not equipped with an epicycle, its orbit being exactly circular. Now, knowing that this circle could not be exactly centred on the Sun, Copernicus resorted to another trick: the centre of the world, the centre of the terrestrial sphere, is not exactly occupied by the real Sun, but by an abstract point, an 'average Sun' that is not much removed. We shall see that this posed a major problem for Kepler, who maintained that the real Sun was the cause of planetary motion.

This model also came up against the same difficulties as that of Ptolemy, since it had to account for the fact that the orbits of the different planets did not exactly fall in the plane of the ecliptic. We shall not go into all the necessary adjustments, which required a mathematical apparatus in the style of Ptolemy. In the end, with its forty or so circles needed, Copernicus' full model[8] quite matched Ptolemy's in complication. Besides, it only indirectly enabled the integration of astronomical observations, obviously made by a terrestrial astronomer, since these had to be noted in relation to the Sun; as a result, it was actually more complicated to use than that of Ptolemy or Tycho Brahe.

Copernicus' motivations are hard to discern,[9] and were not based either on new observations since Ptolemy or on the lack of precision of his geocentric model. His main motivation seems rather to have been theoretical: the absence of rationality in Ptolemy's model,[10] that is, his violations of principles.[11] The following extract from the *Commentariolus*, written around 1513, well sums up the state of the astronomic art at this time, and indicates his motivations in the lines I have italicized:

> Our predecessors assumed, I observe, a large number of celestial spheres mainly for the purpose of explaining the planets' apparent motion by the principle of uniformity. For they thought it altogether absurd that a heavenly body, which is perfectly spherical, should not always move uniformly. By connecting and combining uniform motions in various ways, they had seen, they could make any body appear to move to any position. Callippus and Eudoxus, who tried to achieve this result by means of concentric circles, could not thereby account for all the planetary movements, not merely the apparent revolutions of those bodies but also their ascent, as it seems to us, at some times and descent at others, [a pattern] entirely incompatible with [the principle] of concentricity. Therefore for this purpose it seemed better to employ eccentrics and epicycles, [a system] which most scholars finally accepted. *Yet the widespread [planetary theories], advanced by Ptolemy and most other [astronomers], although consistent with the numerical [data], seemed likewise to present no small difficulty. For these theories were not adequate unless they also conceived certain equalizing circles, which made the planet appear to move at all times with uniform velocity neither on its deferent sphere nor about its own [epicycle's] centre. Hence this sort of notion seemed neither sufficiently absolute nor sufficiently pleasing to the mind. Therefore, having become aware of these [defects], I often considered whether there could perhaps be found a more reasonable arrangement of circles, from which every apparent irregularity would be derived while everything in itself would move uniformly, as is required by the rule of perfect motion.* After I had attacked this very difficult and almost insoluble problem, the suggestion at length came to me how it could be solved with fewer and much more suitable constructions than were formerly put forward, if some postulates (which are called axioms) were granted me.[12]

Copernicus is often viewed as 'the last scholar of the Middle Ages and the first of modern times'. But if he dared to reject Aristotle on the immobility (and centrality) of Earth, his philosophy really was that of a medieval scholar, Aristotelian to the core and, more generally, very respectful of the authority of the ancients, whom he copiously cites in evidence. He believed in the materiality of the orbs pulling the planets, and still based his demonstrations on the intrinsic qualities of objects (nobility, perfection, beauty, etc.): for example, giving in *De revolutionibus orbium coelestium* (I, 10) a justification of the privileged position of the Sun that to us may well seem purely literary or metaphysical:

> At rest, however, in the middle of everything is the Sun. For in this most beautiful temple, who would place this lamp in another or better position than that from which it can light up the whole thing at the same time?

Finally, for reasons that remain somewhat mysterious, he believed that the circular motion of the planets was caused by their spherical shape:

> I shall now recall to mind that the motion of the heavenly bodies is circular, since the motion appropriate to a sphere is rotation in a circle. By this very act the sphere expresses its form as the simplest body, wherein neither beginning or end can be found, nor can the one be distinguished from the other.[13]

We owe to Copernicus:

1. The postulate of Earth's rotation on itself.
2. The introduction of the Sun as centre (but not cause) of Earth's annual orbital motion.
3. The implicit and partial questioning of Aristotle's two physics by giving Earth an ordinary location among the other planets.
4. The premises for a correlation between distances from the Sun and the periods of revolution of the planets.

From the cultural point of view, this de-centring of Earth, and thus of humans, had a bombshell effect, but with a long fuse. In comparison with this, there was a more immediate reaction (in each sense of the term) when Darwin completed this desacralization by linking human origin with that of animals. It is true that in the Renaissance means of communication were less developed. Copernicus was rediscovered nearly a century later, through the work of Kepler and Galileo.

Paradoxically, the author of the model that finally undermined Aristotle's authority did so the better to respect it …

Kepler

As distinct from Copernicus, who remained very discreet about the genesis of his own work, Kepler kept an actual laboratory notebook detailing his advances and setbacks. A great bonus for historians of science! Rarely has a scientist embarked on so many false trails, or described these so bitterly.[14] He actually added above the title of his *Astronomia Nova* the remarkable words *Plurium annorum pertinaci studio elaborate Pragae.*[15]

Kepler sought to go beyond Copernicus' work by finding the *causes* of the planets' movement and of their number.[16] He was convinced of the existence and necessity of a harmony established by divine will. Nothing could be useless or gratuitous, and the instrument of this perfection had to be geometry.

The number of planets

In his first book, *Mysterium Cosmographicum*, published in 1596, Kepler posed the question: why are there precisely six planets?[17] In the wake of his teacher Tycho Brahe, Kepler did not believe in the materiality of the spheres. He first sought to understand their succession by inscribing them in the regular polygons (triangle, square, pentagon, etc.), but without success. He then had the idea of placing himself in three-dimensional space and framing the planetary spheres with polyhedrons. But according to Euclid, there only exist five regular polyhedrons: the tetrahedron, hexahedron (cube), octahedron, dodecahedron and icosahedron. Kepler showed that the spheres of the six planets were arranged in such a manner that the five polyhedrons could be fitted into them: the innermost sphere, that of Mercury, in the octahedron, then that of Venus in the icosahedron, with finally the sphere of Saturn inscribed in the cube and containing the whole. Figure 28 illustrates this construction. Which is why there are only (and for Kepler, can only be) the six planets.[18]

> We see in fact on the basis of propositions 13, 14, 15, 16 and 17 of Book XIII of Euclid's *Elements* that these bodies have by nature the property of being inscribed and circumscribed in this way. Which is why, if these five bodies are each inserted one inside the other, while being separated and enclosed by the orbits, we have the number of six orbits. If there was an age that presented an arrangement of the world of this type, placing six orbits moving around the

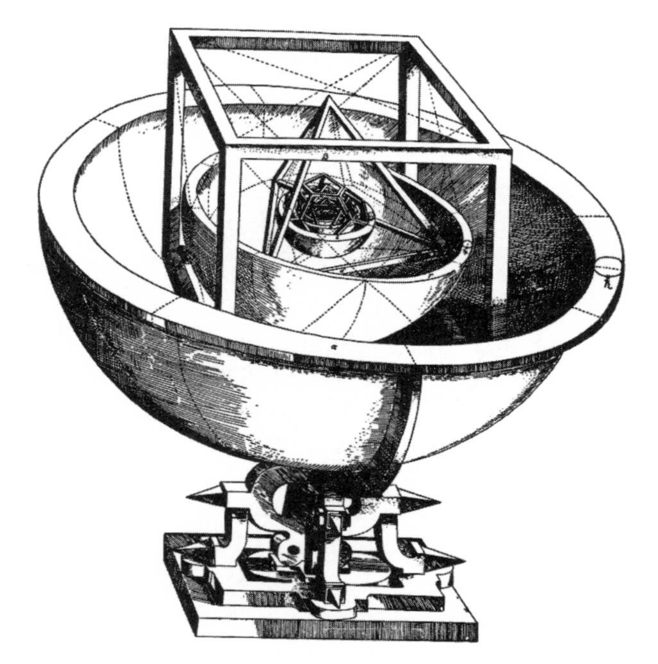

Fig. 28. The projected bowl (financed by the Duke of Württemberg, but never realized) representing Kepler's successive polyhedrons containing the spheres of the planets. The cube (inscribed in the sphere of Saturn) and the tetrahedron (surrounding the sphere of Jupiter) are clearly seen.

motionless Sun, it would certainly be this age that taught genuine astronomy. Now Copernicus offers six orbits of this kind, and places them two by two in such a relationship that these five bodies can all be intercalated as well as possible between these orbits: that is the fundamental idea of what follows.

This model, moreover, not only predicts the number of the planets, but also their respective distances from the Sun.

Very well! Let us finally check whether between Copernicus' orbits we find the relations of these bodies. And to start with, let us look at the matter in the large. The greatest difference between the distances, for Copernicus, is that between Jupiter and Mars, as you see in my presentation of his hypothesis, plate I, and further in chapters XIV and XV. In fact, the distance of Mars from the Sun is not even a third of that between the Sun and Jupiter. Let us therefore seek the body that assumes the greatest difference between the inscribed orbit and the circumscribed one (if we are excused the 'abuse of language' of considering a

> hollow body as a solid): this is the tetrahedron or pyramid. Between Mars and Jupiter there is therefore the pyramid. It is then between Jupiter and Saturn that there is the greatest difference in distance. The orbit of Jupiter, in fact, comes to a little more than half that of Saturn. A similar difference appears between the orbits inscribed and circumscribed by the cube. Saturn accordingly surrounds the cube, while the cube surrounds Jupiter.

To justify this arrangement, Kepler gave a series of arguments based on the particular qualities and affinities of the different polyhedrons. He supplemented this with considerations on a correspondence between the polyhedrons and musical harmonies. It is needless to say that these reasons carry little conviction with modern readers.[19] Kepler himself was not fooled:

> Until now, all that we have said is only some likely indications or probable arguments in favour of the theorem that we have undertaken to study. Let us now move to examine the distances of the orbits of astronomy and to geometrical demonstrations: if all this is not in agreement, then there can be no doubt that the whole of the previous work will have been no more than a pleasant divertissement.

Here we see Kepler faced with the test of truth: do the distances calculated on the basis of this construction agree with those of Copernicus? Assuming a thickness for the (circular) orbs of the planets to account for eccentricities, Kepler obtained the results shown in Table 1.[20] Taking 1,000 as Earth's average distance from the Sun, he used this unit to compare the distances of the five planets from the Sun as per his own polyhedral model with those of Copernicus.

	Kepler	*Copernicus*	*Present-day*
Saturn	9163	9164	9538
Jupiter	5261	5246	5203
Mars	1440	1520	1524
Venus	762	719	723
Mercury	429	360	387

Table 1. Relative distances of the planets from the Sun, when the Earth–Sun distance is taken as 1,000: Kepler's model compared with Copernicus' figures. The third column gives present-day results.[21] While it is true that Copernicus' observational data were better than Kepler's figures, the latter have a different status, being theoretical predictions from his model.

The agreement is not perfect, but it is too good to result from mere chance. What did Kepler do, now convinced of the value of his geometrical considerations? He re-examined the data left by Copernicus. Tycho Brahe's invitation for him to come and work at Uraniborg would play a decisive role, and Kepler himself viewed this as 'the work of divine providence'.

The discovery of ellipses and Kepler's laws

Kepler had, quite rightly, total confidence in Tycho Brahe's results. But despite taking into consideration these new data, he could do nothing to improve the agreement, even with the introduction of epicycles and deferents. He even thought of reintroducing an improved version of the equant, banished by Copernicus, to make his theory agree with Tycho's observations. But in vain.

Here are two examples that show both Kepler's demandingness and his trust in Tycho Brahe's data:

1. An irreducible difference of just 8′ in the orientation of the trajectory of Mars was intolerable for him. Yet 8′ is the angle that a length of 2 millimetres makes at a metre's distance!
2. It is Mercury that has the most elliptical orbit of all the planets known to Kepler. Once again, this deviation from the circle, though far from flagrant, seemed to him unacceptable. And besides, in order to arrive at the ellipse, it was not the study of Mercury that Kepler based himself on, but that of Mars, whose eccentricity is only a fifth of that of Mercury (see Figure 29).

It is impossible to understand this virtual obsession with exactness without knowing that for Kepler the universe had to obey laws of divine geometry, and God does not do approximations. This conviction served him as a principle.

After stubbornly exploring a number of paths, some of them false, Kepler found the solution, thanks to an idea already present in his first book, *The Cosmographic Mystery*: the physical Sun, that is, the real Sun and not an average Sun (the centre of Earth's orbit), is responsible for the movement of all the planets (see next paragraph). This idea came to him from the following double observation: firstly, the planets closest to the Sun have the fastest average speed (as Copernicus had already noted),

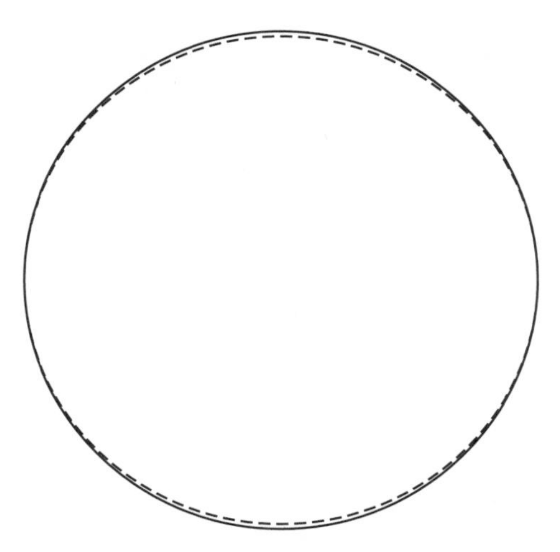

Fig. 29. Comparison of the trajectory of Mercury measured by Tycho Brahe (the dotted ellipse) with a circle of radius equal to half its axis. For the other planets, the difference would be less. Very different from the nice ellipses drawn in school textbooks (and even this book)!

and secondly, their trajectory does not proceed in a uniform motion, but is more rapid when they are closer to the Sun. The physical Sun is thus the cause of their movement; this, and not an immaterial point, is the centre of the world.

How exactly did Kepler proceed? Initially, he continued to accept, after Copernicus, a strictly circular orbit for Earth, but one that was no longer travelled at a uniform pace but according to a 'law of areas' (shadowed on Figure 30): the area swept by the radius joining the true Sun to Earth is proportionate to the time taken to travel it. We see on Figure 30 that the arc B_1B_2 is greater than the arc A_1A_2; and since they are travelled in equal times, the speed increases when the planet approaches the Sun (perihelion). Encouraged by the quality of this hypothesis for Earth, Kepler tried to apply it to Mars. But this was a failure. After a number of attempts, Kepler persuaded himself to replace the circular orbit by an oval; more precisely, by an ellipse, the advantage of which was to specify the role of the Sun, since this was located at one of its foci.

Kepler then understood that Copernicus had been wrong in confining the privilege of being the only planet without an epicycle to Earth. Earth, too, did not follow a circular orbit. Since the observations of the planets were made from Earth, a wrong determination of its orbit introduced a systematic bias into their trajectories. Earth also described an

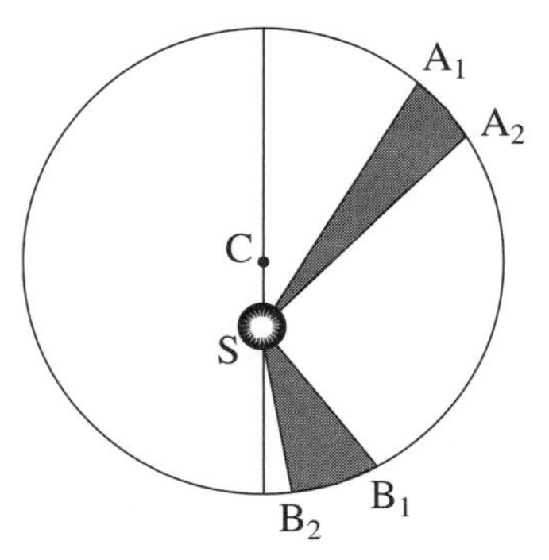

Fig. 30. Illustration of the law of areas. The same interval of time is taken by the movement from A_1 and A_2 as by that between B_1 and B_2. The two shaded areas point to S, the position of the Sun, and not to C, the centre of the orbit, and they have the same area.

ellipse, but of such weak eccentricity that it was possible to equate it with a circle in order to calculate the orbit of Mars without perceptible error.

The way in which Kepler finally arrived at his first two laws is very hard to follow. The result was almost miraculous (Koestler writes of the 'sleepwalkers'[22]). He began from false premises: Kepler believed the speed of the planets to be inversely proportional to their distance from the Sun; he continued for a long time to assume circular trajectories, with an off-centre Sun; and he then had to calculate the areas by an awkward addition of small sections, not having the integral calculus that would come only with Newton and Leibniz.

So much for the story of Kepler's first two laws. The third, which linked the period of revolution to the size of orbit, came only later. Published in 1619, in his voluminous *Harmonices Mundi*, which proposed to illustrate the 'astronomical scales' inherited from Pythagorean theory, it was the most famous result of his work. This third law led from the measurement of the period of a planet, easy to determine, to knowledge of the size of its orbit, a datum difficult to establish from observation. The three laws are summed up in the Glossary below.

And why do they turn?

It is therefore impossible to disassociate Kepler's discovery of elliptical trajectories from the idea that he had of the cause of these motions: the Sun. But what exactly was this 'motive spirit' (later 'motive force') that emanated from the Sun? Kepler was not clear about this, and admitted as much. He proceeded by analogy. Did the Sun not emit something similar to light (or borne by light), whose effect would diminish with the square of the distance? That would explain why the speeds of the planets are weaker according to their distance from the Sun. But this could not be light itself: if that were the case, during an eclipse this action would suddenly stop. A force of a magnetic type? William Gilbert[23] had already shown that Earth was a magnet. But why did the Sun's magnetic power not make Earth fall? Because, Kepler says, the force of the magnet, when exerted on iron filings, does not attract them, but orients them perpendicularly to the radius between them and the magnet. But why does Earth turn? Because the Sun turns on itself, pulling this force-bearing radius that attaches it to Earth. If the Sun's period of rotation is far shorter than the terrestrial year, this is because there are resistances to the motion. The same explanation holds for the rotation of the Moon around Earth. We should note that Kepler, as distinct from Galileo, correctly ascribed responsibility for the tides to the 'tractive force' of the Moon.

Kepler's contribution:

- Kepler's three laws, describing with a previously unknown precision and simplicity the orbits of all the planets, and thereby validating heliocentrism;
- The break with Aristotle's two physics (supralunary and sublunary);
- A step that was still very hesitant, but real, towards the explanation of celestial movements.

Kepler therefore carried out the programme so poetically proclaimed by Giordano Bruno, and that led him to be burned at the stake:

> Cast ridicule on deferent spheres and fixed stars. Shatter and hurl to the ground, with the echo of your vigorous arguments, what blind people see as the adamantine walls of the first mover and last convex. Destroy the central position granted specifically and uniquely to this Earth. Suppress the vulgar belief in the quintessence.

There was however a price to pay for abandoning the idea of two physics. If the stars were not carried by spheres whose supralunary nature it was to be animated (in the etymological sense) by a circular movement, what did carry them? We should not forget that Kepler remained a prisoner to Aristotle's dynamics,[24] clearly summed up by the phrase: '*Quidquid movetur, ab alio movetur.*'[25] The task of maintaining the movement of the planets had to be entrusted to a specific virtue, or soul, or motive force. Kepler did not understand the principle of inertia that Galileo would proclaim, which made such a motor redundant, and Galileo never accepted the elliptical trajectories of Kepler, accusing him of attributing to the Sun a magnetic and even magical role. Ships that pass in the night ...

It was Newton who would bring about a synthesis.

Newton

A general study of what Newton, a giant on the shoulders of giants, brought to science, and even just to astronomy, is beyond the limits of this book.[26]

Newton spent at least as much time studying theology and alchemy as he did on physics and mathematics. The latter studies made him (almost) immortal; the former ones were forgotten.

Newton's centres of interest were varied and often connected: his invention of the infinitesimal calculus, which he called the 'calculation of fluxions', combined with his historical discovery of the universal law of gravity, which we shall return to, enabled him to *deduce* Kepler's three laws. He was interested in optics, where he showed that natural white light breaks down into the prism of primary colours. While developing a particle theory of light, he demonstrated an interference experiment: Newton's rings, explicable only in terms of a wave theory. He invented the telescope, replacing the objective of the astronomical glass by a concave mirror, which brought a tremendous gain in precision for astronomical observations. He was also interested in Earth's shape, showing, against the Cartesians, its flattened shape at the poles.

Newton's personality is disconcerting. Here we have perhaps the greatest scientist of all time showing an obsessive jealousy towards his colleagues: Hooke, with his theory of light and law of universal attraction, and Leibniz, over the paternity of the infinitesimal calculus, to mention only the best-known episodes.

His intellectual world is equally hard to grasp. Keynes, the famous economist, bought at auction in 1936 a trunk containing manuscripts of Newton's that were thought to have disappeared, on alchemy and biblical hermeneutics. Keynes was fascinated by their content, and ten years later published[27] an article entitled 'Newton, the Man':

> I believe that Newton was different from the conventional picture of him. But I do not believe he was less great ... In the eighteenth century and since, Newton came to be thought of as the first and greatest of the modern age of scientists, a rationalist, one who taught us to think on the lines of cold and untinctured reason ... He was the last of the magicians, the last of the Babylonians and Sumerians, the last great mind which looked out on the vision of the intellectual world with the same eyes as those who began to build our intellectual inheritance rather less than 10,000 years ago. Isaac Newton, the posthumous child born with no father on Christmas Day, 1642, was the last wonderchild to whom the Magi could do sincere and appropriate homage.

It is true that we find in Newton a persistent belief in an ancient knowledge, lost or disguised, that it was important to rediscover. This inspired his research in both theology and alchemy. These were ideas characteristic of medieval culture. But Keynes' iconoclastic verdict, which would be more appropriate for Copernicus or Kepler, seems excessive in Newton's case. First of all, because it is unfair to place his biblical exegesis and his alchemical research on the same footing: it is only since the early twentieth century that we understand why research into the transmutation of metals was illusory, given the limited energy sources[28] that the alchemists had at their disposal. So in the seventeenth century there was nothing irrational in researching the transformation of lead into gold: alchemy and chemistry had not yet separated off. Unfair too, since whatever the pertinence of his religious convictions, Newton scarcely ever brings God into his scientific work. And finally, the 'visible world' that Newton contemplated was not at all that of the ancients: the telescope had done its work. Thanks to Galileo, scientists were now familiar with the phases of Venus, the mountains of the Moon, sunspots and the moons of Jupiter; above all, they knew that the Milky Way was in fact 'made up of an innumerable mass of stars'.[29]

In terms of theology, Newton secretly sympathized with the Arian heresy, which rejected the dogma of the Holy Trinity. As president of the Royal Society, and a shareholder in the East India Company, Newton shunned neither honour nor money-making. As director of the Royal Mint, he was pitiless in condemning forgers to death.[30]

In this incomparable scope of achievements, it is only Newton's contribution to the study of the motion of Earth and the planets that we shall discuss in the following pages.

Kepler had empirically established his three laws, and then tried to understand their causes. It is tempting to write that Newton – succoured, indeed, by the results of his predecessor – followed an opposite procedure. In his *Principia*,[31] he proposed a general cause of motion for all the planets – universal gravitation – and then mathematically deduced Kepler's laws from it.

The law of gravity can be summed up as follows: *Two bodies exert on one another a force of attraction directed along the line that joins them. This force is proportional to the product of their masses, and inversely proportional to the square of their distance.*

How to pass from force to motion? Breaking with the whole Aristotelian tradition, and the common sense that underlies it, in modern terms a speed proportional to force, Newton linked force to *change in speed*, that is, acceleration. The fundamental relation of dynamics is: *The acceleration of a body is proportional to the force applied to it and inversely proportional to its mass.*

This means that, in the absence of a force being applied to a body, there cannot be a *change* in its velocity, either in quantity or in direction, but the velocity is maintained. Depending on the initial conditions, the body is therefore either at rest, or in uniform straight-line motion.

These laws are universal and abolish the distinction between a sublunary (corruptible) and a supralunary (perfect) world. That the motion of the planets is (approximately) a circle centred on the Sun means that the velocity of the planets is continuously changing direction, and that they are accordingly subject to a force: the force of gravity. It is this that lies at the root of the 'natural motion' of the sublunary world.

Yet there was a rational kernel to Aristotle's two physics. It is true that, whatever the speed communicated to a ball on a table, it will eventually stop. It is tempting to see a principle involved here, that speed must be maintained by a force in order to last. What was Newton's answer?

Aristotle's sublunary world is one of frictions, those of the air, the table, etc. If the ball eventually stops, it is because it is subject to a force: the force of friction. If, in the imperfect world, friction can be abolished only in thought, it is observable that the more friction is reduced, the farther the ball goes. One can then take the limiting case and maintain that, in the absence of a force being applied, the ball will continue its course indefinitely.

The supralunary world, being perfect, does not have friction. From this point of view, its physics is simpler. It is this that served as Newton's laboratory. He replaced Galileo's thought experiment for establishing the fall of bodies.

The fundamental relationship of Newton's dynamics goes far beyond astronomy; it governs the dynamics of all motion without exception, whatever the forces in play. In particular, it integrates the fall of terrestrial bodies into the general model of attraction of massive bodies.

We shall not dispute the rather cautious custom of asserting that Newton 'explained' the movement of the planets whereas Ptolemy and Copernicus only described them. But what should be said of Kepler? Not only did he describe this motion better, but he attempted to explain it, both by a higher principle of symmetry (the nested regular polyhedrons) and/or by a force of a magnetic type. Newton's theory went altogether deeper. It derived the motion of the planets from simple and universal principles; Kepler's three laws, established empirically, became mathematical consequences of these. In this way, it made possible prediction, which no previous model had done (with the ephemeral exception of Kepler's polyhedrons). The return of Halley's comet was predicted; more generally, the parabolic and hyperbolic trajectories of celestial bodies; and the discovery, by calculation, of a new planet, Neptune.

The discovery of Neptune is emblematic of the triumph of Newton's mechanics. Everything began with the observation that the trajectory of Uranus (found by William Herschel in 1781) was slightly askew from the ellipse predicted by theory. Either the theory had to be modified or, if it were maintained, the existence postulated of a still unknown planet whose gravitational effect was disturbing the trajectory of Uranus. Urbain Le Verrier sent the result of his calculation to the astronomer Johann Galle at the Berlin observatory. On 23 September 1846, pointing his telescope very close to the place indicated at the time given, he discovered the new planet. As Arago would elegantly tell the Académie des Sciences: 'Monsieur Le Verrier saw the new planet at the end of his pen.' The problem was ripe: John Adams in England had carried out a similar calculation but, unfortunately for his future fame, had failed to publish it.

Emboldened by this success, Le Verrier tried to explain the anomalies in the orbit of Mercury in the same fashion, but in vain. In this case, there really was a violation of Newton's laws. This would eventually be understood with the theory of general relativity. This theory, however, was not developed *in the wake* of observation, any more than the theory of special relativity was developed *in the wake* of Michelson's experiment that showed the invariance of the speed of light. No more had Copernicus constructed his model following the inaccuracies of Ptolemy's. In each of these three cases, it was the internal critique of the previous theory that provided the spur. Verification came later. Of course, this mechanism for the birth of a new theory cannot be generalized.

The world did not stop turning after Newton. We need to make clear the context of this theory. As we shall show below (p. 143) on the subject of the Coriolis force, it is important to define carefully, in speaking of velocity or acceleration, the frame of reference against which this takes place. Moreover, it was only by making this frame of reference clear that direct measurements of Earth's motion were established, and only in the nineteenth century.

But if the key to explaining the motions of the planets is the law of gravity, what explains this law? What mechanism can ensure instantaneous action at a distance? Kepler ventured to offer an answer, followed by Descartes with his vortices (a precursor of Maxwell's vortices two hundred years later, to explain the propagation of electromagnetic waves). The theory of relativity dispensed with the concept of force, taking Einstein's equations as its point of departure. But do these equations follow from more general principles? The chain seems endless. We shall just say that a further notch has been gained in the depth of explanation. Let us try to show in what sense this is the case.

The Cradle of Modern Physics

Study of the place and motion of our planet had not just brought responses to the questions of astronomy that it involved. It originated the modern scientific conception of the world, which is broadly based on mathematical physics. The great physicist Eugène Wigner (1902–95) neatly referred to 'the unreasonable effectiveness of mathematics':[32] since Galileo maintained that the world 'was written in the language of mathematics', there has been a colossal and continuing literature on this subject.[33] But the non-scientific reader need not share my own amazement at seeing objects being born initially as solutions to equations before they exist as subjects of observation.

Let us give three examples among many of the 'unreasonable effectiveness' of mathematical equations:

- Hertz discovered in 1887 the radio waves predicted in 1864 by Maxwell's equations;
- Anderson discovered in 1932 the electrons with positive charge (antimatter) predicted in 1928 by Dirac's equations;
- Observations from the satellite Uhuru in 1975 convinced the scientific community of the existence of black holes, as predicted by the equations of relativity theory more than half a century earlier.

It is commonplace today to emphasize the impact of the computer revolution on everyday life, but the eruption of mathematics into the sciences, strongly stimulated by the study of Earth, had a far greater importance. In order to grasp this, we have to abandon the naïve conception of mathematics as simply a calculation tool, a conception that is unfortunately very widespread even in the scientific milieu. When a well-known French sociologist of science writes: 'Although this is indeed a decisive advantage provided by what is aptly called an equation (because it ties different things together and makes them equivalent), this advantage should not be exaggerated',[34] one might think that this is applicable even to the mathematics of Ptolemy or Copernicus. But the equations of Newton (or Einstein) are of a quite different order. It is false for Latour to add:

> First, an equation is no different in nature from all the other tools that allow elements to be brought together, mobilized, arrayed and displayed; no different from a table, a questionnaire, a list, a graph, a collection; it is simply, as the end-point of a long cascade, a means to accelerate the mobility of the traces still further.

This primitive conception of mathematics is similarly upheld by a former minister of national education:

> Mathematics is in the process of being devalued, in an almost inevitable way. Now there are machines to do calculations. The same for the construction of curves.[35]

An equation, and in particular a differential equation,[36] is not so much an 'end-point' that sums up existing knowledge, as a starting point: Newton's equations (just as all fundamental equations of modern physics) are in fact research programmes. An example less well known than the discovery of Neptune, but certainly much more profound, is the theory of deterministic chaos. This theory, which establishes the difference between determinism (the existence of law) and the ability to predict beyond a certain horizon, has profound cultural implications. It is there embryonically in Newton's equations, when these seek to describe, for example, the motion of a system of three planets.

What is deterministic chaos?

The regularity of Earth's motion is such that it served for a long while as the standard of time: the second was defined as the 86,400th part of an average terrestrial solar year. If the universe consisted only of Earth and the Sun, taken as abstract points (approximation 0), this regularity, mathematically demonstrable by Newton's equations, would be verified over an infinite time. But the solar system is far more complex than this, made up of all the planets, satellites and asteroids. With the greatly simplifying hypothesis that all celestial bodies are point-sized, it is still easy to express in equations the motions of all the planets. But we have no way of finding explicit solutions to this system of coupled equations (coupled, since the planets all interact on one another). The natural way is to start from the solution with approximation 0 and see the presence of other bodies as small perturbations. Earth's trajectory, for example, then appears as an ellipse corrected by a series of successive terms, the effects of which, it is hoped, are increasingly negligible. In this way, Laplace obtained a very good improvement in calculating planetary trajectories, valid for hundreds of thousands of years (there was no idea at that time of the age of the solar system). He enthusiastically wrote:

> All events, even those which on account of their insignificance do not seem to follow the great laws of nature, are a result of it just as necessarily as the revolutions of the sun. In ignorance of the ties which unite such events to the entire system of the universe, they have been made to depend upon final causes or upon hazard, according as they occur and are repeated with regularity, or appear without regard to order; but these imaginary causes have gradually receded with the widening bounds of knowledge and disappear entirely before

> sound philosophy, which sees in them only the expression of our ignorance of the true causes ... We ought then to regard the present state of the universe as the effect of its anterior state and as the cause of the one which is to follow. Given for one instant an intelligence which could comprehend all the forces by which nature is animated and the respective situation of the beings who compose it, an intelligence sufficiently vast to submit these data to analysis, it would embrace in the same formulae the movements of the greatest bodies of the universe and those of the lightest atom; for it, nothing would be uncertain and the future, as the past, would be present to its eyes.[37]

Alas! Rigorous examination of the influence of successive terms of the theory shows that, after a certain rank, this influence increases instead of declining. Which means that beyond a certain time, we can no longer be certain of our predictions. Mathematics, with its 'unreasonable effectiveness', proves that this method is no longer reliable on the scale of hundreds of millions of years. One might believe that computers today would make it possible to resolve the system of equations numerically and dispense with analytic series that do not converge, that, starting from the initial conditions, one could follow the trajectory by using Newton's equations. But nothing of the kind! Poincaré, in the late nineteenth century, showed in fact that, as soon as more than two bodies are in play, this system may display an extreme sensitivity to initial conditions: a very small change in these eventually induces a tremendous change in the determination of the trajectory:

> A very small cause,[38] which escapes us, determines a considerable effect which we cannot but notice, and then we say that this effect is due to chance. If we knew exactly the laws of nature and the situation of the universe at each instant, we would be able to predict exactly the situation of this same universe at a later moment. But, even when the laws of nature no longer hold any secrets for us, we will only be able to know the situation approximately. If this enables us to predict the subsequent situation with the same approximation, that is all that we need, and we say that the phenomenon has been predicted, that it is governed by laws; but this is not always the case, it may happen that small differences in the initial conditions generate very great ones in the final phenomena; a small error in the former producing a tremendous error in the latter. Prediction then becomes impossible and we have a chance phenomenon.[39]

Now, when a physical system is sensitive to small changes in the initial conditions or the environment, to the point that its subsequent evolution is unknowable, the calculations that describe this evolution are equally unstable. This is simply because a computer operates with a limited number of decimals, and this limitation[40] plays the same role as a minuscule change in the initial conditions.[41] For this type of system, a calculation using ten decimal places would give long-term predictions radically different from those from a calculation with twenty decimal places, and the predictions of both these two calculations would have nothing in common with those that would have been obtained from the mathematical solution whose existence mathematicians had proved. These theoretical predictions, therefore, are forever inaccessible, which is not too serious, since the sensitivity of the system to the inevitable modifications of the environment deprives them of any interest.

What Poincaré described, and what today is called deterministic chaos, is that in certain situations (cases of chaotic systems such as the solar system) even a perfect knowledge of the laws of development does not make it possible to predict beyond an extremely limited horizon.[42] It has to be understood that this inability to predict is not due to either complexity or ignorance of the laws, in which case it would be trivial.

Laplace has often been accused of a naïve determinism. First of all, however, the passage cited above is taken from a book devoted to probabilities; and secondly, Laplace was very careful to write: 'an intelligence which could comprehend all the forces by which nature is animated and the respective situation of the beings who compose it'. He clearly did not suspect the devastating effect that the lack of precision in knowledge of initial conditions would have. Poincaré did not contradict Laplace's determinism; he simply marked, in certain cases, the limits to prediction.

The future rational mechanics was first of all verified in astronomy, with the movement of the planets, and not on Earth, despite the greater part of its applications being here. We owe it to Newton to have launched what Wigner called the unreasonable efficacy of mathematics in the natural sciences. This would be an open door for subsequent scientific discoveries in many fields.

SOURCES

First of all, it is necessary to consult, if not read, authors in the original text. For Copernicus, Kepler[43] and Newton, this is far from easy – no comparison with the clarity of Galileo! In Copernicus and Kepler there is an inextricable mixture of medieval thinking and extraordinarily modern ideas. With Newton we are talking more of a juxtaposition of these two worlds: his famous *hypotheses non fingo*[44] meant that, in his scientific texts, Newton hated the introduction of metaphysical considerations or those external to the subject; loyal to his principles, he did not practise this mixture of genres. For all of them – including Newton, who did not use the integral and differential calculus despite being one of its inventors – the reasoning is always geometric and difficult to follow.[45] The help of the classic historians of science is indispensable; we have drawn principally on Delambre, Pannekoek, Koestler,[46] Koyré, Neugebauer, Simon and Verdet.[47]

Koestler's work deserves special mention. The author's great literary talent makes us relive the atmosphere of the great period of discoveries. He particularly shares with us his admiration, even worship, of Kepler's personality, perseverance and talent. He shows the exceptional role that Kepler played as an individual in the history of science, and how Galileo, despite his rejection of the 'religion of Aristotle', could not accept the discovery of elliptical trajectories. Yet the criticisms of J.-P. Verdet, O. Gingerich and J.-P. Luminet rightly draw attention to Koestler's hasty judgements on Ptolemy: 'the work of a pedant with much patience and little originality, doggedly piling "orb in orb"'; Copernicus: 'a stuffy pedant, without the flair, the sleepwalking intuition of the original genius'; and Galileo, whose 'self-importance' he insistently denounces. More serious to my mind, he continued (in 1955), despite his impressive breadth of culture, to perpetuate a tradition that is nowadays largely rejected[48] of a total medieval obscurantism with nothing, or even a regression, between Aristotle and Copernicus. He repeats the idea that 'Europe knew less geometry in the fifteenth century than in Archimedes' time'. The explanation for this is allegedly 'the division between reason and faith' (see Part One of his book, chapter 4). Koestler denounces 'the tenets of the age of scientific materialism which begins with Galileo and ends with the totalitarian state and the hydrogen bomb' (*sic*). This explanation is deplorable: Einstein responsible for Hiroshima? At all events, such a divorce did not exist – or scarcely so – in the Middle Ages. On the contrary, one might say, both Christians and Muslims were convinced that every advance in knowledge would make us more appreciate the greatness of the Lord, the wisdom of his Creation and the rightness of the Holy Book. The divorce in question began in the Renaissance, when this communion lost its self-evident character in the wake of scientific progress, and it was not synonymous with regression – very far from this. Besides (but is it 'besides'?), Koestler gives the impression of flirting with a naïve conception of the 'conservatism of modern science' when he writes (ibid.):

> Thus for the last thirty years, an impressive body of evidence has been assembled under strict laboratory conditions which suggests that the mind might receive stimuli emanating from persons or objects without the intermediary of the sensory organs, and that in controlled experiments, these phenomena occur with a statistical frequency that invites scientific investigation.[49]

Why, he asks, do scientists accept the paradoxes of quantum mechanics and reject the transmission of thoughts? But this conservatism of scientists that Koestler denounces does not explain anything: it did not lead them to reject quantum mechanics, whose predictions are still more unreasonable. We shall come back to this point in the conclusion (p. 152).

On the other hand, Koestler is right to insist that good reasons can often lead to a wrong judgement, and vice versa. Galileo, for example, (wrongly) rejected the (right) ellipses of Kepler on account of the latter's (actually) hazy considerations on magnetism. One of the reasons that drove Darwin to (rightly) extend Kelvin's calculation of the age of Earth was his (wrong) conception that 'Nature does not make leaps'. Pasteur, in his (justified) opposition to Pouchet over spontaneous generation, was strongly sustained by his religious convictions. And so on. We see very well, then, the burgeoning development of science on which natural selection operates, thus avoiding a naïve history which would just be that of the winners.

There are many more modern and very educational books on Renaissance astronomy, in French particularly the works of Depondt and Véricourt, Sénéchal, Blamont and above all Gapaillard, which is both very concise and complete.[50] And for its clarity and precision, the Spanish book by Rioja and Ordóñez.[51]

CHAPTER 6

Distances

The Greeks were able at a quite early date to estimate precisely the *relative* sizes and distances of the planets, but they could only very poorly estimate their absolute values. It was not until the eighteenth century that it was possible to assess the Earth–Sun distance correctly, and with the aid of Kepler's laws this then made it possible to measure the dimensions of the orbits of all the planets.

Measuring the distance *d* from a point A to an inaccessible object C can be done by the method of parallax, or triangulation. You measure the distance from A to an accessible point B, and AB is called the *base* of the triangulation. You then measure the two angles $\alpha = C\hat{A}B$ and $\beta = C\hat{B}A$. Point C is then determined. Figure 31 illustrates a difficulty inherent to astronomy: if *d* is very large in relation to the base, then angles α and β are very close to 90°, the straight lines AC and BC almost parallel, so that the intersection point C is very far away and poorly determined. In this case, a minimal error in measuring the angles can lead to an enormous error in distances. With the help of a little trigonometry, the diagrammatic construction, which is very imprecise when the base is small, is unnecessary: knowledge of AB and of the two lines of sight makes it possible to calculate *d*. But this was not possible before Hipparchus and the first trigonometrical tables.

The permanent problem in astronomy is then to find bases that are 'only' (!) some hundreds of thousands of times smaller than the distances to be measured. A terrestrial base is typically in the order of a few thousand kilometres; it is impossible to use this for measuring the distance to the stars, even the closest ones, which are some 40,000 billion kilometres away. Using the positions of Earth at an interval of six months gives a base of 300 million kilometres. This makes the measurement of distance to the nearest stars just possible, given instruments of great precision that were not available in Galileo's time. We recall (p. 81) that the absence of parallax was one of the major scientific arguments

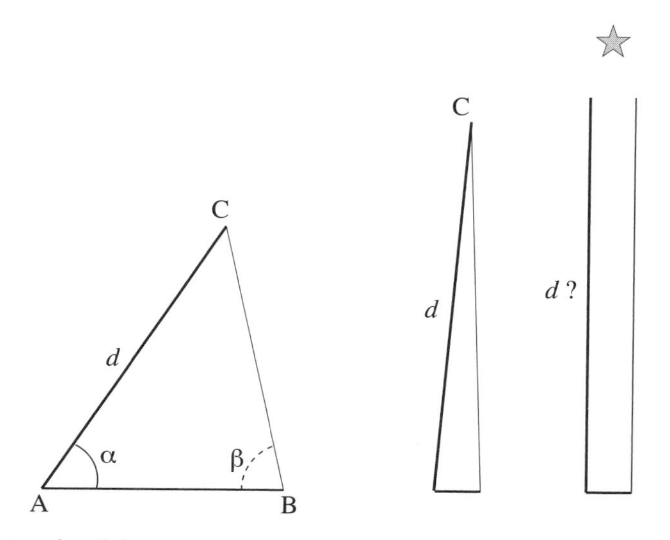

Fig. 31. Principle of the method of triangulation. Let *d* be the distance between a terrestrial point A and a celestial object C. This can be determined by measuring (or calculating) the distance between A and another known point B, and the lines of sight of C from these two points (see text). The middle figure shows what happens when C is very distant. Finally, the right-hand figure gives, to scale, what we obtain by measuring the parallax of the nearest star with the biggest possible terrestrial base.

opposed to the movement of Earth. The calculated segment AC is often used as a new base for determining farther distances. But it is necessary to avoid the accumulation of errors. That is why the logical progression goes from knowing Earth's size to knowing that of the Earth–Moon distance, then Earth–Sun and finally the distance to the nearest stars.

The Moon, Its Radius and Its Distance From Earth

The Earth–Moon distance was measured by triangulation for the first time in 1751, on a Berlin–Cape Town base (more or less on the same meridian) by Lalande and La Caille. This could not have been done before a precise measurement of the radius of Earth was achieved, or before the astronomical telescope could determine angles with sufficient precision. Today, thanks to the laser reflector positioned on the Moon, we know this distance to within a centimetre, even better than that from Paris to Lyon. But how did the ancients proceed?

The first astronomer who proposed a method of measuring was perhaps Aristarchus of Samos, mentioned above as the ephemeral pioneer of the heliocentric model. The only text of his that has come down to us[1] is entitled 'On the sizes and distances of the Sun and the Moon'. Aristarchus was the first to have the idea of using a lunar eclipse to deduce the Moon's size as a function of that of Earth, subsequently its distance from Earth, and finally the Earth–Sun distance. It is remarkable to note that Aristarchus' distances are all wrong, despite the methods he used being quite correct. How was that possible? His geometry was irreproachable, but his observations lacked precision.

Aristarchus noted that in the course of a lunar eclipse, the Moon passes through Earth's shadow (Figure 32). He began by measuring the time taken by the Moon to move by one lunar diameter in relation to the background stars, which is about 1 hour. He then measured the duration of eclipses, about 2 hours, and deduced from this that 'the breadth of the shadow is two Moons'. If the Sun was at an infinite distance, the shape of this shadow would be cylindrical, and the Moon's radius R_M would be half that of Earth R_E. But Aristarchus took account of the fact that the shadow is conical, and by clever geometrical arguments obtained the result that $R_E \approx 3R_M$, which was not at all bad (the modern value is 3.67). Knowing now the diameter of the Moon and the angle it makes as seen from Earth (apparent diameter), he deduced its distance, as explained in Figure 33. The apparent diameter of the Sun being the same as that of the Moon, Aristarchus estimated with this method the diameter of the

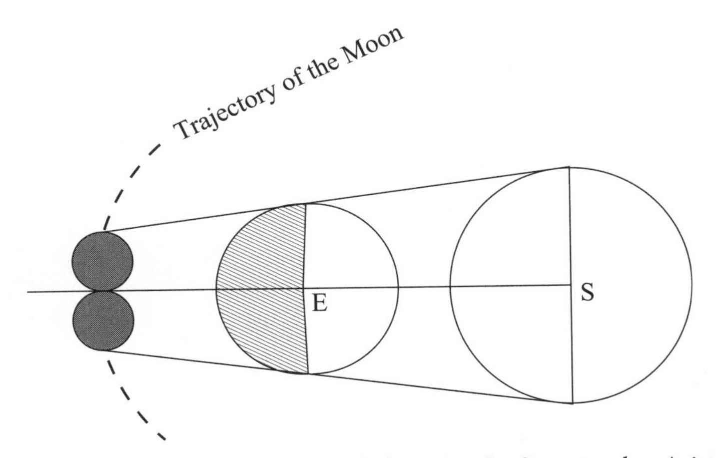

Fig. 32. Principle of the estimation of the Moon's diameter by Aristarchus. Obviously not to scale.

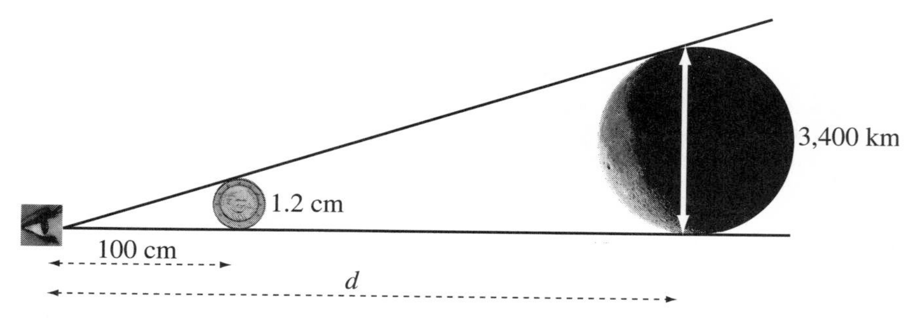

Fig. 33. Aristarchus' principle for estimating the Earth–Moon distance. With the data of Hipparchus, the Earth–Moon distance *d* stands in the same relation to the distance of the coin (100 cms) as the Moon's diameter (3,400 kms) to the coin's diametre (1.2 cms), so $d = 100/1.2 \times 3400 \approx 283{,}000$ kms.

Sun on the basis of its distance from Earth (see following paragraph). But Aristarchus' numerical data were faulty. Rather remarkably, for example, he took an apparent diameter of 2° instead of 0.5°, which placed the Moon far too close. Two centuries later, but on the same principle, Hipparchus improved the calculation a good deal and gave values for the Earth–Moon distance and the diameter of the Moon that are quite close to those established today.

The Sun, Its Radius and Its Distance From Earth

To determine the Earth–Sun distance, Aristarchus began by remarking that the Moon 'receives its light from the Sun'. In other words, it does not constitute a source of light, but only reflects sunlight. Then he turned his attention to the case when the Moon is exactly half lit ('dichotomous' Moon). In this case (Figure 34), the Sun–Moon direction is necessarily perpendicular to the Earth–Moon direction. Aristarchus actually used a parallax method with the Earth–Moon distance as its base; he measured the other line of sight and found $\beta = 87°$. Unfortunately, he did not have trigonometry, which would immediately have given him the ratio of the distances Earth–Moon and Earth–Sun $EM/ES = \cos \beta$; but by subtle reasoning he showed that $18 < ES/EM < 20$, not far from the 19.1 given by exact calculation. His estimate for β, however, which is very hard to make, was bad. In actual fact, $\beta \approx 89°85'$. This may seem a small error, but it corresponds to $ES/EM \approx 380$, or an error of a factor of 20 in the estimate of this ratio! Since the EM distance was also poorly estimated,

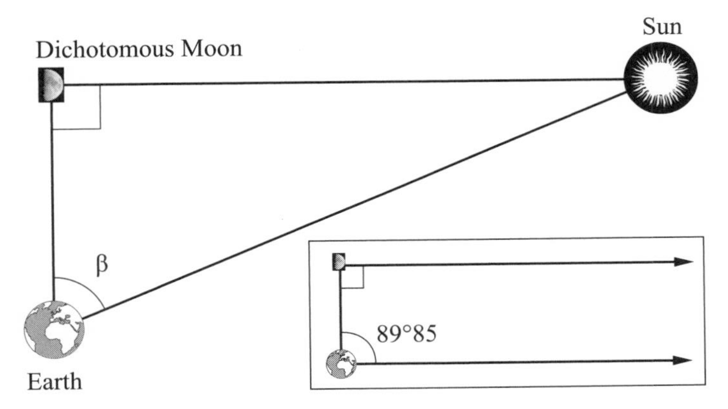

Fig. 34. Principle of Aristarchus' estimate of the Earth–Moon distance. In the inset, the drawing is to scale. It is clear that a small error in the angle β, which is actually very close to a right angle, leads to a colossal error for the Earth–Sun distance.

as we saw, it is understandable how the ancient world was mistaken on the question of the Earth–Sun distance, generally underestimated by a factor of 80.

The first 'modern' estimates of the Earth–Sun distance date from the seventeenth century. Cassini began in 1672 by measuring the Earth–Mars distance by triangulation, taking a base from Paris to Cayenne in Guyana; then, using the parameters of the ellipse described by Mars, and Kepler's third law, he deduced from this for the first time a good order of magnitude for Earth's distance from the Sun. We should note that to 'measure' the distance to the Sun, Cassini used Kepler's laws, as Rømer did to estimate the speed of light.[2] In the following century, Halley (1656–1742) proposed a different and more precise method of parallax, and in 1761 the transit of Venus across the Sun was used. This measurement possibly marked the first major international scientific cooperation.[3]

The Distances of the Planets

The distances of the planets from the Sun have been determined very precisely by the parallax method, using the diameter of Earth's orbit as a base. The Newtonian theory of gravity alone did not make it possible to explain the values of the orbital radii (shown in Table 2). Prediction of the planets' positions, as well as their masses, actually assumed an extremely detailed model of the formation of the solar system.

n	*Planet*	*Measured as*	*Titius-Bode*
$-\infty$	Mercury	0.39	0.4
0	Venus	0.72	0.7
1	Earth	1	1
2	Mars	1.52	1.6
3	(Ceres)	2.77	2.8
4	Jupiter	5.2	5.2
5	Saturn	9.54	10
6	Uranus	19.2	19.6
7	Neptune	30.1	38.8

Table 2. Relative distances of the planets from the Sun, with the Earth–Sun distance taken as 1. Comparison between the Titius-Bode law in the right-hand column and the observed values in the adjacent column.

In the eighteenth century, the astronomers Johan Daniel Tietz or Titius (1729–96) and Johan Elert Bode (1747–1826) proposed an empirical relationship that approximately accounted for the sequence of orbital radii of the six planets known at the time. This sequence, in Astronomical Units, is of the form

$$d = 0.4 + 0.3 \times 2^n, \text{ where } n = -\infty, 0, 1, 2, 3, 4, 5$$

for Mercury, Venus, Earth, Mars, an unknown planet, Jupiter and Saturn respectively. In the eighteenth century, this relationship aroused a great deal of interest, and it was even thought possible to make up for the insufficiency of the theory of gravitation and embark on a theory that would go beyond that of Newton. This 'law' seemed validated by the discovery in 1781 of Uranus (which would be $n = 6$) and the asteroid Ceres in 1801, which filled the gap for $n = 3$. There followed a series of discoveries of small bodies at distances from the Sun comparable to that of Ceres, hence the hypothesis that these were fragments of a planet that had disintegrated. But a detailed analysis of the orbits of these bodies (which form what is now called the asteroid belt) refuted this hypothesis. Besides, the sum of the mass of all these bodies did not even correspond to that of a planet less massive than Mercury. The decisive blow to the Titius-Bode law, however, was the discovery of Neptune in 1846. The planet envisaged for $n = 7$ was far more distant from the Sun than Neptune actually was. Despite this, the law is still often mentioned as being true but unexplained, and is greatly prized by astrologers.

This distribution of planetary distances could be explained by a two-stage process. Four and a half billion years ago, the planets were formed by the aggregation of matter from a disc gravitating around the Sun. Their masses and positions were the result of fluctuations in the density of this disc. At the end of this stage, which lasted only a few million years, the gaseous planets were closer to the Sun than they are today. Subsequently, the gravitational interaction between the planets reciprocally modified their orbits, until they finally reached the present equilibrium position. Recent calculations of the dynamics of the solar system show that there are resonance mechanisms that create stable (or quasi-stable) orbital zones.[4] It cannot be completely ruled out, therefore, that a relationship of the Titius-Bode type might exist, resulting from very complex dynamic processes.[5] I see it as more probable, however, that its status is that of Kepler's nested regular polyhedrons, which he used to determine the orbits of the planets: simply a lucky chance.

CHAPTER 7

The Battle over Heliocentrism

> It took Newton to see the Moon falling, when everyone could see clearly that it didn't fall.
>
> – Paul Valéry, *Mélanges*

So far we have been examining models of Earth simply from the standpoint of physics. How did society receive these?

It is a delicate matter to separate what in modern terms could be called the scientific aspects of the quarrel from the religious ones. Galileo, for example, inspired by the long tradition of biblical commentators, sought to cast off the constraint that a literal reading of the Bible imposed. His procedure is not only of historical interest: in an age when various fundamentalisms are again on the rise, we find, written nearly four hundred years ago, a line of argument that I see as inescapable for any believer in one of the great monotheist religions (Christian, Jew, Muslim), if he or she does not want to ignore the results of science.

The crucial question of knowing whether Earth *really* is immobile and the centre of the universe lay at the root of the battle between the Catholic church and Galileo. Like many historical battles, it has often been simplified into a stereotyped image: a backward and uneducated mass opposed to a misunderstood and solitary genius embodying progress.

That is a simplistic view, since to a large extent the struggle was waged within the church itself, according to its universally accepted rules of the game: the Bible as divine word, and therefore unchallengeable. Until the eighteenth century, scientists (intellectuals) were all believers, even men of the church, if with differing responsibilities. Besides, Galileo was far from being isolated or misunderstood. Just like Kepler, with whom he corresponded, he was honoured as one of the leading mathematicians of his time. He was for a long

while the protégé of various prelates, in particular Maffeo Barberini, the future Pope Urban VIII, who nonetheless had him condemned in 1633.

The critique of this naïve view, however, should not lead us to forget that this marked a historical turning-point in the manner of conceiving knowledge. Galileo explains in a famous passage in *The Assayer* (1623):

> **Philosophy is written in this grand book, the universe, which stands continually open to our gaze. But the book cannot be understood unless one first learns to comprehend the language and read the letters in which it is composed. It is written in the language of mathematics.**

Even if Galileo claimed otherwise, this was a rupture with the dominant idea that the 'book of nature' was written in the Bible, which scholars had to interpret (to justify). Galileo indeed makes fun of those who think 'to seek truth neither in the world nor in nature, but (I cite their own words) in the comparison of texts' (letter to Kepler, 19 August 1610).

We have seen how the hypothesis according to which Earth is animated by a double movement of rotation, on itself in 24 hours and around the Sun in a year, had already been proposed by Aristarchus of Samos in the third century BCE. But the crass violation of common sense that this primitive heliocentric model implies, added to its lack of precision, made it of little usefulness.[1] It was forgotten to the benefit of geocentrism, more natural since it envisaged the world from the standpoint of an inhabitant of Earth.

Ptolemy's system, despite being extremely complicated, was *natural* in the sense that:

- it accounted satisfactorily for everyday experience (the Sun rises in the east and sets in the west), and for astronomical observations made up to the sixteenth century;
- it obeyed the Aristotelian aesthetic (inherited from Pythagoras and Plato) which imposed circular trajectories on the planets;
- finally, and this was the most important point in the age of the Counter-Reformation, it conformed to the Bible, which saw Earth as immobile. Here is the striking passage from the book of Joshua that 'proves' that the Sun is in motion:

> (10:12) On the day when the LORD delivered the Amorites into the hands of Israel, Joshua spoke with the LORD, and he said in the presence of Israel:
>
> Stand still, O Sun, in Gibeon;
> Stand, Moon, in the Vale of Aijalon.
>
> (10:13) So the sun stood still and the moon halted until a nation had taken vengeance on its enemies, as indeed is written in the Book of the Just. The sun stayed in mid heaven and made no haste to set for almost a whole day.

This text played a determining role in the condemnation of Galileo in 1633.[2]

Just Before Galileo

Giordano Bruno and Tycho Brahe were totally different individuals and moved on totally different paths. The first was a mystic obsessed by a vision of the cosmos, the second a meticulous observer of the positions of the heavenly bodies. Both helped to shatter Ptolemy's construction.

Giordano Bruno

Bruno was a Dominican monk, born at Nola, close to Naples, in 1548. He died in 1600, burned alive in Rome by the Inquisition, with his tongue nailed down to prevent him from speaking. He had undergone a seven-year trial, in the course of which he was tortured on several occasions. Among other blasphemies and heresies that questioned the Trinity and the Immaculate Conception, he championed a plurality of inhabited worlds and the system of Copernicus. But he was not in fact a Copernican, in the sense that for him the universe was infinite and could not have a centre. Refusing to abjure, he has often been seen as a martyr to science.[3] Bruno was less of a scientist than a visionary philosopher,[4] but he did nonetheless have a certain prescience of the idea of gravitation. In his second dialogue, he wrote:

> This is why, if we say and mean that the movement of these things proceeds up and down, this is understood for a certain region and from a certain point of view; with the result that if something moving away from us is heading towards the Moon, then we say that this thing is rising, while those who are on the Moon, our 'anticephalids', will say it is falling.[5]

The lyricism of his fifth dialogue is moving. He has his spokesman Albertino say to Filoteo, his opponent:

> Reveal to us the true substance, matter, act and efficient cause of the whole, and how every sensible and composite object is built up from the same origins and elements. Convince our minds of the infinite universe. Rend in pieces the concave and convex surfaces which would limit and separate so many elements and heavens. Pour ridicule on deferent orbs and on fixed stars. Break and hurl to earth with the resounding whirlwind of lively reason those fantasies of the blind and vulgar herd, the adamantine walls of the *primum mobile* and the ultimate sphere. Dissolve the notion that our earth is unique and central to the whole. Remove the ignoble belief in that fifth essence. Give to us knowledge that the composition of our own star and world is even as that of as many other stars and worlds as we can see. Each of the infinity of great and vast worlds, each of the infinity of lesser worlds, is equally sustained and nourished afresh through the succession of his well ordered phases. Rid us of those external motive forces together with the limiting bounds of heaven. Open wide to us the gate through which ether is even as that of our own world. Make us clearly perceive that the motion of all of them proceedeth from [the impulse of] the inward soul; to the end that illumined by such contemplation we may proceed with surer steps toward a knowledge of nature.

The episode of the statue in honour of Giordano Bruno deserves to be told, being so closely tied up with the political history of Italy.[6] Erected in 1849 after the ephemeral triumph of the Roman republic, on the execution site of the Campo di Fiori, it was destroyed by Pope Pius IX. In 1876, profiting from the vagaries of the birth of the Italian state, an international committee that included in particular Victor Hugo, Ernest Renan and Mikhail Bakunin proposed a subscription to reconstruct it, leading to violent confrontations rather like the Dreyfus affair in France. In 1889 it was finally in place, despite the threat of Pope Leo XIII to abandon Rome if the statue were erected. In 1929, with the Lateran agreements designed to normalize relations between the Fascist regime and the Vatican, Pius XI's hopes were raised and he requested the destruction of the statue; but Mussolini refused,[7] asserting on 13 May 1929:

> I have to declare that the statue of Giordano Bruno, melancholy as was the fate of this monk, will remain where it is.

Tycho Brahe

Seduced by the simplicity of Copernicus' model, but frightened by its radicalism which, contrary to the Scriptures, no longer placed Earth (and man) at the centre of the world, Tycho Brahe (1546–1601), the celebrated Danish astronomer, proposed a compromise, as we said above: Earth was indeed immobile at the centre of the universe, the Sun turned around it, but *surrounded by all the other planets.* Besides its compatibility with the Bible, this model presented the advantage of an excellent predictive power.[8] This made it nicely acceptable to the Jesuits. But it remained artificial.

Galileo

Galileo Galilei was totally convinced[9] that the Copernican model was correct by the discovery that he made (as early as 1610) of the moons of Jupiter (politically christened the Medicean stars in honour of his illustrious protector Cosimo II de Medici).

Galileo did not immediately come into conflict with the church. Did not St Thomas Aquinas (1225–74) make the following observation?

> Thus in astronomy the theory of eccentrics and epicycles is accepted (observed phenomena) being given that what appears to the senses as the movements of the stars is respected by this hypothesis; this is not however a decisive proof (that they are the true cause of these phenomena), as it is not said that another hypothesis would not respect them as well.[10]

The publication of Copernicus' theory, *On the Revolutions of the Heavenly Spheres*, was cautiously preceded by an introduction, 'To the reader concerning the hypotheses of this work',[11] which presented this theory as a mathematical model to simplify the description of celestial movements, and not as reality.[12] We read here:

> For it is the duty of an astronomer to compose the history of the celestial motions through careful and expert study. Then he must conceive and devise the causes of these motions or hypotheses about them. Since he cannot in any way attain to the true causes, he will adopt whatever suppositions enable the motions to be computed correctly from the principles of geometry for the future as well as for the past.[13]

As has often been said, it is rare for such an important book to have passed unnoticed.[14] We do not even know whether Luther had read it when he wrote his condemnation of heliocentrism.[15] The Catholic church, in contrast to Protestant strictness,[16] initially appeared more benevolent. Copernicus' services were even sought for the establishment of the new calendar. But things took a bad turn when Galileo was no longer content to use the heliocentric model to save the phenomena, but claimed that Earth really did turn. The ban of 1616, which declared the teaching of the Copernican doctrine to be contrary to the Scriptures, did not single out Galileo personally, but his claim to a truth contradicting the Bible. The teaching of the said doctrine was now banned, since it

> is foolish and absurd, and false in philosophy, and formally heretical, as it explicitly contradicts, and at several points, the dicta of Holy Scripture, read according to the proper sense of the words and the common interpretation of the Holy Fathers and theologians.

Solicitous to show that the movement of Earth was more than a 'mathematical hypothesis', Galileo believed he had found a proof of it in the phenomenon of the tides: the speed of a mass of water on Earth results from Earth's two composite speeds: its rotation on itself in 24 hours, and its annual rotation around the Sun. According to the time (or position), these speeds either counter or augment each other. In sufficiently large oceans, therefore, there are differentials of speed that generate movements of the water. As we know today, this theory is wrong.[17] It even contradicts Galileo's own theory, according to which the movement of Earth cannot be shown by observations made from itself. Kepler already had the right intuition that the tides were caused by the influences of the Moon and Sun (high tides when these act in the same direction, low tides when they are opposed). Galileo knew that Kepler believed in the occult effects of the heavenly bodies, but saw an explanation of this kind as pertaining to magic. In the fourth day of his *Dialogue Concerning the Two World Systems*, he has Salviati say:

> [Kepler], however, leant his ear and gave his assent to an empire of the Moon on the water, of occult properties and other childishness in the same vein.[18]

It is amusing to note that Kepler, rightly suspected of believing in astrology, had a better intuition here than the scientific Galileo, who railed at his credulity.[19]

Newton gave the beginnings of a correct explanation[20] of the tides in terms of the law of universal gravitation. But at the time, it was Galileo's argument that prevailed,[21] including with the future pope, Cardinal Barberini (see below, p. 134). Still to show the rotation of Earth, Galileo advanced on top of the tides the argument of the trade winds. He was right, but by intuition, as he could not know the Coriolis effect (see Appendix A) that is the real cause of these. As for heliocentrism, Galileo judged that Copernicus had developed a model of the solar system that was simpler and more elegant than that of Ptolemy, while being at least as precise, and thanks to his telescope he gathered observations that confirmed this. It was now known that:

- Venus, just like the Moon, has its phases, which Ptolemy's model could not explain;
- Jupiter has satellites.

This last point was not strictly speaking material evidence, but it did accredit the idea that the condition of satellite (small bodies orbiting around larger ones) was not unique: not only was there the Moon orbiting Earth, and the five other planets orbiting the Sun, but even moons around Jupiter.

Given the poor quality of his instrument, far inferior to the smallest modern binoculars, Galileo's gifts of observation demand admiration. The chromatic distortions of his instrument, which were considerable, provided arguments for his detractors:[22] they maintained that the image given by the glass did not reflect reality. The Moon, for example, could only be a perfect sphere according to Aristotle, and Jupiter's satellites or the beads of Saturn[23] were an optical illusion.

Galileo and the Weight of the Bible

Galileo's thinking was very modern, compared to other scientists of his day – or even a century and more later. Each time that reason and experience seem to contradict Holy Writ, he says that it is wrong to interpret this literally, without understanding that it was designed for the ignorant people of the time. The 'errors' that people believe they find in it actually belong, as it were, to a divine pedagogy adapted to the days of Moses; it is the duty of modern scientists not to remain imprisoned by the Scriptures, even at the risk of seeming to ridicule them.

A captivating biography of Galileo is given by Festa,[24] and a detailed account of his trial and its context in Boriaud.[25] The following quotations are taken from Galileo's letter to Benedetto Castelli of 1613 and his 'Considerations on the Copernican Opinion' of 1615.

1) The literal interpretation can lead to blasphemy:

> I should have added only that, though the Scripture cannot err, nevertheless some of its interpreters and expositors can sometimes err in various ways. One of these would be very serious and very frequent, namely to want to limit oneself always to the literal meaning of the words; for there would thus emerge not only various contradictions but also serious heresies and blasphemies, and it would be necessary to attribute to God feet, hands and eyes, as well as bodily and human feelings like anger, regret, hate and sometimes even forgetfulness of things past and ignorance of future ones. Thus in the Scripture one finds many propositions which look different from the truth if one goes by the literal meaning of the words but which are expressed in this manner to accommodate the incapacity of common people; likewise, for the few who deserve to be distinguished from the masses, it is necessary that wise interpreters produce their true meaning and indicate the particular reasons why they have been expressed by means of such words. (Letter to Castelli)

2) Must one start from the Bible?

> Now, one wants to know where to begin in order to ascertain its falsity [i.e., that of the Copernican system], that is, whether from the authority of Scripture or from the refutation of the demonstrations and observations of philosophers and astronomers. I answer that one must start from the place which is safest and least likely to bring about a scandal; this means beginning with physical and mathematical arguments. For if the reasons proving the earth's motion are found fallacious. (Considerations on the Copernican Opinion)

In that case, for our Florentine scholar, it would be proved that reading of the Holy Book was sufficient unto itself, and everything was fine,

> But if those reasons are found true and necessary, this will not bring any harm to the authority of Scripture.

The conclusion that imposes itself is:

> Therefore, beginning with the arguments is safe in any case.

This is how Galileo solves the problem, without ever denying that

> the Holy Scripture can never lie or err, and that its declarations are absolutely and inviolably true. (Letter to Castelli)

On the other hand, the words of the Church Fathers are historically dated:

> I am always told that, in interpreting the passages of Scripture relevant to this point, all Fathers agree to the meaning which is simplest and corresponds to the literal meaning; hence, presumably, it is improper to give them another meaning or to change the common interpretation, because this would amount to accusing the Fathers of carelessness or negligence. I answer by admitting that the Fathers indeed deserve reasonable and proper respect, but I add that we have an excuse for them very near at hand: it is that on this subject they never interpreted Scripture differently from the literal meaning, because at their time the opinion of the earth's motion was totally buried and no one even talked about it, let alone wrote about it or maintained it. (Considerations on the Copernican Opinion)

As far as the 'salvation of the soul' is concerned, since it is scarcely probable that any experiment could ever wrong-foot the Holy Book, Galileo deems it wiser to keep to the literal sense:

> Because of this, it would be most advisable not to add anything beyond necessity to the articles concerning salvation and the definition of the Faith, which are firm enough that there is no danger of any valid and effective doctrine ever rising against them. (Letter to Castelli)

Is this a matter of prudence, or a concession with the help of an assertion that does not commit him to anything?

Although written before his trial, these texts are structured as legal defences. The distinction is carefully made between 'texts dictated by God' that are unchallengeable and that one is forced to interpret, and those of the Fathers of the Church, who were certainly 'inspired by God', but victims of the prejudices of their time and accessible to criticism.

In a less developed but more imaginative manner, Kepler championed a similar position:

> No doubt the word of God is something important, but the finger of God is equally so. And who would deny that the word of God was adapted to its purpose and, because of this, to the ordinary way of talking of men? Consequently, to distort the word of God in things absolutely evident so that it refutes the finger of God in nature is something that any truly religious person will do all that they can to avoid.[26]

A Rich Tradition of Exegesis

Though he does not explicitly refer to it, Galileo was brought up in the culture of the Middle Ages, even if this was far less 'medieval' than Renaissance writers liked to maintain.[27] What marked a hardening in the positions of the Catholic church was the Counter-Reformation,[28] formalized at the Council of Trent,[29] against the advance of Protestantism. A few examples will give an idea of the soil that nourished Galileo.

More than ten centuries earlier, St Augustine had declared[30] that there were 'allegories in Scripture'.

In the thirteenth century, St Thomas Aquinas explained[31] that taking Scripture too literally risked 'turning it to ridicule'.

St Bonaventura (1217–74) maintained in the thirteenth century that

> Scripture being addressed to simple and poor people, often spoke in the common fashion. And so when it spoke of heaven, it spoke as if heaven were apparent to our senses.[32]

At the same time, in canto IV of *Paradiso*, Dante wrote:

> For this reason the Scripture condescends to your capacity, and attributes feet and hands to God, while meaning otherwise; and Holy Church represents to you with human aspect Gabriel and Michael.

In the fourteenth century, Nicole Oresme (1325–82) apparently argued the rotation of Earth on itself.[33] He observed that the Holy Scriptures contain passages where God is angered and then calms down like a common mortal. Can these passages be taken literally? For the same reason, one should not take literally what is said in the Bible about the rotation of the Sun.[34] Further on, however, in the tradition of scholastic *disputatio*, he presents contrary arguments:

> Besides, although Averroes says in ch. XXII that movement is more noble than rest, the contrary is evident, as even according to Aristotle in the same ch. XXII, the most noble thing that there is or can be is in its perfection without movement; and that is Good. Beside, rest is the final cause of movement; which is why, according to Aristotle, the bodies here below are in movement towards their natural place in order to be at rest there. On top of this – and this is the sign that rest has higher value – we pray for the dead that God give them rest: *Requiem aeternam*, etc.

This quotation is quite emblematic of the difficulty that we have today in understanding certain reasonings of the thinkers of this time. Reference to masters (Averroes, Aristotle) serves as demonstration.[35]

This manner of using sacred texts actually has a long history.[36] In the Islamic lands, two philosopher-doctors played a predominant role: Avicenna (Ibn Sina) (980–1037) and above all Averroes (Ibn Rushd) (1126–98). The latter, an acclaimed commentator on Aristotle for the Christian West, held a position in relation to reading the Koran that was close to that of Galileo, for example:

> We say still more: there is no assertion of revelation whose obvious sense is in contradiction with the results of demonstrations, without it being possible to find, by proceeding to the inductive examination of the totality of particular statements of the revealed text, another statement whose obvious sense confirms the interpretation or is near to confirming it.[37]

Does this mean that if need be, and provided that a sufficiently vague assertion confirms it, it is possible to make a *sura* say the contrary of, or at least something other than, what it seems to say? The subsequent influence of Averroes, like that of many other 'philosophers', was almost nothing in the Islamic world, and certainly not comparable with that which he enjoyed in the Catholic church.

In the Jewish world, Saadia Gaon (882–942), the highest rabbinical authority of his time, and the predecessor of Maimonides, developed a quite original point of view. For him, science (the laws of reason) and sacred text (the revealed laws) could not contradict one another;[38] but the prescriptions of the latter, even if hard to understand, prevailed with a double advantage. First of all, they enabled time to be gained for reaching the truth:

> If the doctrines of religion can be discovered by rational inquiry and speculation, as God has told us, how can it be reconciled with His wisdom that He announced them to us by way of prophetic Revelation and verified them by proofs and signs of a visible character, and not by rational arguments? To this we will give a complete answer with the help of God. We say: God knew in His wisdom that the final propositions which result from the labour of speculation can only be attained in a certain measure of time … Perhaps many of us would never have completed the work because of their inability and never have finished their labour because of their lack of patience; or doubts may have come upon them, and confused and bewildered their minds. From all these troubles God (be He exalted and glorified) saved us quickly by sending us His Messenger.

Second, they made access possible 'even to women, children, and people incapable of speculation'. We should remember that, faced with the same line of questioning, Galileo's answer was quite different: 'Consequently, to begin with rational arguments is a sure path in all cases.'

Moshe ben Maimon, known to us as Maimonides (1138–1204), a Jewish sage, also a medical doctor and philosophical contemporary of Averroes, put forward a view quite close to that of Galileo – but four hundred years earlier – in his famous *Guide for the Perplexed*:

> They [the scholars] have warned you that these things mentioned are obscure … To speak of all these subjects, homonymous words have been used, so that the common run of men can take them in a certain sense according to the measure of their intelligence and the weakness of their understanding, and that the perfect man who has received instruction can take them in a different sense.[39]

In the same work, he writes still more clearly:

> the incorporeal nature of God has been demonstrated, and it is necessary to have recourse to the allegorical interpretation, on each occasion that the literal sense is refuted by a demonstration; we then know in advance that it is necessarily subject to interpretation.

Maimonides played a similar role to that of his Islamic compeers for Christian thought in the Middle Ages.

And finally, we should signal the famous heretic Spinoza, who wrote five centuries later in a similar vein, as shown in the following passage:

> No one will be inconvenienced, on the least reflection, that to have a true idea of faith it is necessary to know that the Scripture was not appropriate only for the intelligence of prophets, but was also put within reach of the Jewish people, the most variable and inconstant that has ever been. Whoever, in fact, takes indifferently everything in the Scriptures for a universal and absolute doctrine of Divinity, and does not careful discern from all the rest what was appropriate for the common intelligence, must necessarily confuse the opinions of the people with heavenly doctrine, take the fictions and concerns of men for divine teaching, and abuse the authority of the Scripture.[40]

The possible interpretation of the Book was a matter for scholars, and for them alone. Such was the definitive and unanimous view of these thinkers, who condemned the very idea of popularizing an allegorical version intended for people with little instruction, who would risk suddenly losing their faith. Sometimes even the use of allegory was forbidden for those under twenty-five years of age,[41] just as certain films today are banned to minors under eighteen. In case of difficulty in the interpretation of the sacred text, the authority of the elders (rabbis, *ulemas*, or Fathers of the Church) provided the only mainstay of demonstration. This tradition lasted broadly until the seventeenth century, and even today it remains popular in certain milieus. Peter Godman writes:

> The mistake, however, could be avoided by obedience to the elite represented by the Inquisition and the Congregation of the Index, which in Bellarmine's lifetime forbade translations of the Bible into Italian. This subversive book was in fact read by the 'simple-minded and ignorant' (*simplices et idiotae*) whom the Holy Office was supposed to protect. It accordingly prevented their straying by proscribing the word of God in the vernacular tongue and offering this

> to the faithful in a Latin that they did not understand.[42] It followed, by a circular but effective reasoning, that they needed a priest.[43]

The present relationship that believers have to the Koran is rather similar: first of all, the great majority of Muslims do not speak Arabic; second, the language of the Koran is hard even for the mass of Arabic speakers to understand. There is, of course, the recurrent problem of the interpretation of the text, but on top of this there is a problem of language, even if Arabic,[44] stabilized by the study of the Koran, has evolved less in thirteen centuries than either French or English.

There is a continuous spectrum between the simple faith of hundreds of thousands of followers of the Abrahamic religions and a sophisticated view such as that of Spinoza or Galileo, excluding anthropocentrism and being the privilege of intellectuals. This spectrum is crucial: without the faith of the former, the latter would not exist, or would play only a marginal role. The tradition of all the great monotheist religions is to restrict this second approach to enlightened milieus alone, whether de facto or de jure. The attempts of Benedict XVI or the guardians of Islamic power to stifle the 'modernizations' of their respective churches are fuelled by the same tradition. How does this spectrum come about? It goes without saying that, for all believers, it is the poetic charm of the sacred texts that establishes the connection between ingenuous readers and scholars. After all, ties based on imagination or shared sentiments are at least as effective as those based on cold reason.

The Relationships of Scientists to the Bible

Though seen as one of the greatest scientific geniuses of all times, Newton remained in matters of religious philosophy a man of the Middle Ages; at the very least, his position was ambiguous. He did indeed recognize:

> As to Moses I do not think his description of ye creation either Philosophical or feigned, but that he described realities in a language artificially adapted to ye sense of ye vulgar. Thus where he speaks of two great lights, I suppose he means their apparent, not real greatness. So when he tells us God placed those lights in ye firmament, he speaks I suppose of their apparent not of their real place, his business being not to correct the vulgar notions in matters philosophical but to adapt a description of ye creation as handsomly as he could to ye sense & capacity of ye vulgar.[45]

We have seen (p. 126) how Newton recalibrated the six days of Creation. Moreover, the same Newton maintained rather astonishingly

> That religion & Philosophy are to be preserved distinct. We are not to introduce divine revelations into Philosophy, nor philosophical opinions into religion.[46]

The allegorical reading of the Bible, in other words, should be reduced to a minimum, and by and large Newton saw Holy Scripture as a source of unchallengeable knowledge; one should

> acquiesce in that sense of any portion of Scripture as the true one which results most freely & naturally from the use & propriety of the Language & tenor of the context in that & all other places of Scripture to that sense. For if this be not the true sense, then is the true sense uncertain, & no man can attain to any certainty in the knowledge of it. Which is to make the scriptures no certain rule of faith, & so to reflect upon the spirit of God who dictated it.
>
> He that without better grounds then his private opinion or the opinion of any human authority whatsoever shall turn scripture from the plain meaning to an Allegory or to any other less naturall sense declares thereby that he reposes more trust in his own imaginations or in that human authority then in the Scripture. And therefore the opinion of such men how numerous soever they be, is not to be regarded. Hence it is & not from any reall uncertainty in the Scripture that Commentators have so distorted it; And this hath been the door through which all Heresies have crept in & turned out the ancient faith.[47]

Without going into the general evolution of relations between scientists and organized religion, let us mention that Descartes, impressed by the condemnation of Galileo, decided not to publish his book *The World* (also known as *Treatise on the Light*), which championed heliocentrism.[48] And even a hundred years later, Leibniz bent himself into intellectual contortions to remain politically correct.[49]

The end of the Aristotelian conception of motion abolished the need for a 'prime mover', which was perhaps, along with the existence of miracles, the main rational justification of God in the medieval period. In the eighteenth and nineteenth centuries, it was more the marvellous arrangement of the cosmos, and in particular the extraordinary complexity of life, that kept this justification alive. In the nineteenth century, as we saw with Kelvin, the idea that a watchmaker[50] regulated, or at least had regulated, the universe, was very widespread. As for today's 'intelligent design', this is simply a remix with contemporary flavour of the 'cosmic destiny' that Bertrand Russell polemicized against, not without humour, in his *Science and Religion*:[51] a conception that was very fashionable after the First World War, its French representative being Henri Bergson (1859–1941), who had a strong influence on Teilhard de Chardin (1881–1955). As for Benedict XVI, in his answer[52] to the question 'how to bring science and faith into harmony', that 'behind everything there is a great intelligence, in which we can trust', is not very far from 'intelligent design' (see note 60, p. 217, and note 53, p. 235).

Finally, a Catholic view of Genesis is expressed very well by a great scientist, the Belgian Georges Lemaître (1894–1966), an ordained priest and discoverer of what he called the 'primeval atom', later rechristened by Fred Hoyle ironically as the 'big bang'. Lemaître's view was relatively authorized, as he was president of the Pontifical Academy of Sciences. His God is a hidden one, and at all events not to be called upon in scientific discussion. In 1936, Lemaître declared:

> In a certain sense the scientist brackets out his faith in his research, not because this faith could encumber him, but because it has nothing directly to do with his scientific activity. Likewise a Christian does not behave differently from a non-believer when it is a matter of walking, running or eating.[53]

The affinity with the quote from Newton (p. 126 above) is evident.

To sum up a certain evolution of the church very schematically, it is possible to distinguish a rational church based on Aristotle,[54] at least from the thirteenth century, then a church that was anti-rationalist,[55] 'anti-scientific' until the nineteenth century, and finally a present-day church[56] able even to claim the work of scientists such as Stengers or Prigogine, and quantum mechanics, as justifying the unintelligibility of this world below by science alone. Formerly, no contradiction was seen between knowledge of nature (physics) and the Book: truth was one, everything contributed to the admiration of the work of the Lord. Sadly, however, the progress of science would drive a wedge through this nice agreement, provoking a struggle against 'blinkered science'. But, God be praised, salvation would come from the sciences themselves when they ventured – apparently, at least – to question determinism. All the same, this did not prevent Benedict XVI from taking a certain step back when he denounced, in October 2008, 'the problems and risks of modern exegesis', with the consequence that 'those who do pure exegesis risk seeing the Bible as a book of the past'.

The Catholic religion, by virtue of being relatively universal both socially and geographically, is perhaps more subject than others to the intellectual pressures of the secular world.[57] Moreover, since it is based on a well-established hierarchy and discipline (popes, cardinals and bishops deriving their legitimacy from St Peter), it has both felt the need and had the means, over time, to afford the luxury of a broad interpretation of the Sacred Book. Minois recounts the rifts in the church on this point.[58]

The *Catechism of the Catholic Church* presented by Pope John Paul II in 1992 casts an interesting light on this. Those who have only the Book as cement, Protestants, Jews and Muslims, have been more timid in their interpretations, at least at the level of public declarations. Is this for fear of schism?[59]

The Other Great Monotheist Religions

In contrast to Catholics, neither Muslims, Jews nor Protestants have a hierarchy or clergy deemed the bearers of a unified and universal word, which makes it that much harder to follow their developments.

Kepler was a Protestant. As against Newton (and in part also Galileo), he never dissociated his faith from his scientific work. How did he manage to reconcile his biblical reading with his Copernicanism? He wrote:

> Although it is in conformity with piety to examine, at the start of this dispute over Nature, whether anything is said that is contrary to the Holy Scriptures, I believe it none the less inopportune to embark on this controversy here, before I am forced to. I promise, however, in general, to say nothing that does injury to the Holy Scriptures, and if it happens that Copernicus and myself should be convinced of such a thing, to hold [what he says] as void.[60]

Advised by good friends who were more politic than he was, Kepler did not take this argument further, and moved immediately on to a defence of the Copernican system.[61]

We should point out that in statements officially attributed to Luther we find the following accusation against an anonymous opponent (Copernicus is not named):

> People gave ear to an upstart astrologer who strove to show that the earth revolves, not the heavens or the firmament, the sun and the moon. Whoever wishes to appear clever must devise some new system, which of all systems is of course the very best. This fool wishes to reverse the entire science of astronomy; but sacred scripture tells us that Joshua commanded the sun to stand still, and not the earth.[62]

If it is true that the Protestants condemned heliocentrism earlier and more strongly, and hence indirectly pressed the Vatican to do likewise (via the Council of Trent), and if it is also true that they never formally revoked this condemnation, one should in no way underestimate their practical rationalism: in the seventeenth, eighteenth and nineteenth centuries, it was a plethora of Protestant (and Anglican) scientists who led the way in the development of new scientific theories.

The Jews, as is well known, had a very rich tradition of reading the Torah:[63] this book manifestly contained the truth, but as we do not know how to grasp this, we must constantly learn to read it. The work of doctors, scholars, exegetists and other commentators became decisive. As with the hard sciences, this was a cumulative process. The Bible would also be inconceivable without the Talmud.[64] If, from a philosophical point of view, the reading grids of the sacred text are very close in the cases of Maimonides, a Jew, and Averroes, a Muslim, the imprint that each left is totally different. The tradition of Maimonides continues today, while that of Averroes in the contemporary Islamic world is very weak. A book by Colette Sirat gives a detailed insight into Jewish medieval philosophy.[65]

The Koran, which presents by and large the same genesis as the Bible, is thus faced with the same problems.[66] This is true for both the dating of Earth and the emergence of man from clay. An interpretation which could almost be called 'Galilean' (in the sense of a text that was certainly inspired by God, but in such a way that ignorant people of the time could understand it and that scholars now have the task of interpreting it) flourished in the ninth and tenth centuries. Its best-known representatives were Avicenna and Averroes. But following the defeat of the Mu'tazilite[67] movement, this tradition was severely repressed. This was the end of the golden age of the Arabic-Islamic world. The beginnings of a modern Islamic reformism date only from the late nineteenth century, particularly in Egypt,[68] its best-known representatives being perhaps Jamâl ad-Dîn al-Afghânî (1838–97) and Mohammed Abduh

(1849–1905), who revived the tradition of a humanly created and thus more broadly interpretable Koran.

Both the Bible and the Koran are rather long texts, but it is striking to note that the respective religions published a mass of authorized texts many times greater than their sacred books with a view to teaching their reading (the Talmud and the writings of the Church Fathers in the one case, the Sunna in the other).[69] In the end, commentators played an essential role, given the ambiguity of the texts. It would be tempting[70] to pursue the comparison with certain philosophical texts that gladly confuse profundity with obscurity.

It is a given fact today that the Bible (and likewise the Koran) is neither a book of astronomy nor a book of history. If this conclusion was established for science in the nineteenth century, as far a great majority of natural scientists are concerned, it is relatively recent in relation to history.[71] None the less, the Bible does contain, perhaps in passing, assertions on both history and natural science. This raises no problem for atheists: it simply presents the best of the beliefs, knowledge and customs of its time. For them, we venture to say, there are no miracles. For believers, however, the Sacred Book, inspired (if not actually written or dictated) by God, 'cannot lie'; it is impossible to see it as no more than a dated book that reflects the culture of a particular age. There remains then the following alternative:

- Either the Book contains the truth, but we are unable to see this, hence an *apparent* contradiction between science and faith, which the necessarily very copious commentaries have the task of explaining;
- Or what scientists hold true today will be false tomorrow: for example, the age of Earth or the theory of evolution. Science is fragile, but not the revealed text.

Two points of view that in fact are often combined.

Traditionally, believers in a revealed truth, biblical or koranic, have had to confront the sacred texts with the findings of science. Since God is capable of anything except contradiction, exegetists have most commonly applied themselves to a delicate balancing work between the Sacred Book and the book of nature. Galileo's originality was to have broken with these equivocations and mocked those who sought truth 'in the confrontation of texts', maintaining that it was necessary to start 'with natural and mathematical reasons'. In short, Galileo, as distinct from all others, 1) distinguished assiduously between assertions that are false but made for pedagogic ends, dictated by a God who is himself infallible, and error due to the ignorance of the doctors of the faith who, as men of their time, perpetuated an immediate reading of the Sacred Book. And 2), he did not hesitate to contradict this or that verse of the Bible without the support of another verse or the authority of a recognized dignitary (Aristotle or a Church Father).

Rehabilitation of Galileo or of the Church?

There is a naïve image[72] of Galileo's condemnation, and also one of his rehabilitation. Here are some major dates for reference:

1600 Giordano Bruno burned at the stake.

1616 Papal condemnation of heliocentrism.

1633 Condemnation of Galileo.

1741 'Faced with optical proof of the orbiting Earth',[73] Benedict XIV granted the *imprimatur* for the first edition of Galileo's complete works, followed in the nineteenth century by the successive lifting of prohibitions (which was clearly not the same thing as approval).[74] But there was still resistance within the church, and it was not until the new Index was published in 1846 that all bans on the presentation of Copernican ideas were lifted (ibid.).

1930 Canonization of Cardinal Bellarmine, subsequently awarded the title of doctor of the church.[75] In the report of over five hundred pages drawn up by Cardinal Caietano Bisleti,[76] which lists the services that this eminent prelate rendered to the church, only a few lines mention Galileo and Giordano Bruno. Much use is made of Pierre Duhem (1861–1916)[77] in order to criticize Galileo as a physicist:

> Logic was on the side of Osiander, Bellarmine and Urban VIII, and not that of Kepler and Galileo; the former had understood the exact import of the experimental method, and in this respect, the latter had been confused.

> The history of the sciences, on the other hand, celebrates Kepler and Galileo, whom it places in the rank of reformers of the experimental method, while it does not pronounce the names of Osiander, Bellarmine and Urban VIII. Is this sovereign justice on its part? … The Copernicans stubbornly held to an illogical realism, despite everything leading them to drop this error, and by attributing to astronomical hypotheses the correct value that so many authorized men had determined, it was easy for them to avoid both the quarrels of philosophers and the censure of theologians.

In the official text proclaiming Bellarmine's doctorate, Cardinal Pacelli,[78] the future Pius XII, never mentions Copernicus, Bruno or Galileo. On the other hand, vibrant homage is paid to such holy men as Bellarmine, described as following Torquemada as a 'hammer of the heretics … by whose work the attacks with which the heretics threatened these same truths [i.e., those of faith] were disposed of'.

1979 On 15 December, under the title 'Theology Is Ecclesiastical Science', John-Paul II explained at the Gregorian University of Rome:

> And if we must recognize that scientists were not exempt from the cultural conditionings of their milieu, we can also note that precursors of genius were not lacking, such as Saint Robert Bellarmine in the case of Galileo, who wished useless tensions and damaging conflicts to be avoided in the relations between faith and science.

1992 31 October saw John-Paul II's famous speech to the Pontifical Academy of Sciences. The problem was not so much to rehabilitate Galileo as to plaster the wounds of a painful past:

> A tragic and reciprocal incomprehension has been interpreted as a constitutive opposition between science and faith. The elucidations brought by recent historical studies permit us to maintain that this painful misunderstanding now belongs to the past.

What incomprehension? What misunderstanding? The main protagonists perfectly understood one another but did not agree. That is all. The word 'misunderstanding' makes it possible to avoid that of disagreement, and thus to avoid deciding who was right and who was wrong, who enabled science to advance and who held it back.

A little remark on anachronism is needed here. The defenders of Galileo were accused of basing themselves on work 'more than a century later' to validate assertions that were not validly proved by the scientist, for example, the argument from tides. At the time, therefore, it was the church that was right. This is what Cardinal Ratzinger indirectly maintained in 1990, drawing on the philosopher Paul Feyerabend. The future Pope Benedict XVI offers here an 'a fortiori' argument, since, he tells us, this philosopher, 'despite being a sceptic and agnostic', wrote:

> The church at the time of Galileo was much more faithful to reason than Galileo himself, and also took into consideration the ethical and social consequences of Galileo's doctrine. Its verdict against Galileo was rational and just, and revisionism can be legitimized solely for motives of political opportunism.

This assertion of a 'rational and just' verdict is completely faithful to the position of the church as expressed on the canonization of Bellarmine in 1930 (see above). It is true, moreover, that Galileo was not tortured, but only shown the instruments to be used. Everyone has their own idea of what is 'rational and just'! And this detail of torture, or at all events the threat of it, while carefully forgotten by Pope John-Paul II, was not so by Galileo,[79] though his perspective was of course far more subjective:

> But then, if one knows oneself so much superior to the opponent without leaving the domain of the natural and without using other weapons than these philosophers, why then have recourse to confrontation with a terrible weapon, impossible to escape, and that terrorizes by its sight alone the cleverest and most experienced champion?

This was a long way indeed from a drawing-room conversation between gentlemen trying to convince one another simply by rational argument.

If anachronism is rejected, it must be rejected completely: in its condemnation, both of heliocentrism in 1616 and of Galileo in 1633, the church did not base itself on the falseness of the evidence of Earth's orbiting. It always stood on the same ground, that of the 'primacy of the Scripture as interpreted by the Holy Fathers'. Here is how Cardinal Agostino Oregio, the future papal theologian, relates an interview (perhaps from 1615)[80] between Galileo and Barberini:[81]

> Concessis enim omnibus, quae vir doctissimus excogitaverat, quaesivit, an potuerit an sciverit Deus alio modo disponere orbes vel sidera, ita ut

quaecunque vel in caelis apparent phaenamena, vel de siderum motibus, ordine, situ, distantia ac disposione dicuntur, salvari possint. Quod si negas, sanctissimus dixit, probare debes implicare contradictionem posse haec aliter fieri, quam excogitasti. Deus enim infinita sua potentia potest, quidquid non implicat contradictionem: cumque Dei scientia non sit minor potentia, si potuisse Deum concedimus, et scivisse etiam affirmare debemus. Quod si potuit ac novit Deus haec alio modo disponere, quam excogitatum est, ita salventur omnia, quae dicta sunt: non ad hunc modum debemus divinam arctare potentiam et scientiam. Quibus auditis quievit vir ille doctissimus. Ex quo et ingenii et morum laudem retulit.[82]

In fact, after having accepted [for a moment] all the propositions that this very great scholar had conceived, he asked [him] whether God could have disposed the spheres, or would have been able to dispose them, or the stars if you like, in a different manner, with the result that, whatever these things are that appear in the heavens and are called astronomical phenomena, what is said of their movements, succession, position, distance and disposition is saved. [But] if you maintain, declared the very holy prelate, that these phenomena cannot be produced in another manner than you have imagined you must recognize that this would imply a contradiction. In fact, by his infinite power God can do all that does not imply a contradiction, and as the science of God is no less than his power, if we accept that God could have done it, we must also accept that God would have been able to [do so]. And if God was able and capable to dispose these astronomic phenomena in a different manner than that which we have imagined, [yet] with the result that all these astronomical phenomena that have been mentioned are saved, [then] we must not reduce the divine power and knowledge to the manner [that you have imagined]. After having heard these words, the very great scholar remained silent. It is for this reason that the very holy prelate praised [his] intelligence and conduct.

In other words, since natural philosophy alone cannot account for Earth's immobility, another logic must be appealed to: what presumption to try and confine divine omnipotence within the laws of physics pronounced by men!

It is instructive to read the sentence of condemnation in full.[83] Nowhere does it mention any mistake of scientific reasoning on Galileo's part. What it does emphasize, on the other hand, is the danger of questioning the authority of the church, at least as much so as that of the sacred text. That is what the Vatican's defenders at the time called 'Galileo's pride', which was equally responsible in the end for his condemnation.[84]

Poincaré's (2001) opinion on this question, six years before Duhem, is given below on p. 145.

If it is hard to understand the Vatican's position, this is because the papal declarations on Galileo, as on all other subjects, claim to be continuous:[85] each pope refers positively to the previous. How can one venerate what one almost burned? What we thus see is more in the way of gradual slippages. There is an apparent evolution of the church, as long as we ignore certain popes. In clear contradiction with Urban VIII (in 1633), Pope John-Paul II (in 1992) appreciated the theologian in Galileo but not the physicist, since Galileo was unable in his day to prove the movement of Earth, as Bellarmine had so correctly maintained. It is this (re)reading of Galileo's trial that is anachronistic.

The church was not opposed to Galileo on the basis of his errors in physics, but rather because he defended the primacy of what we today call the scientific approach over obscurantism, i.e., an approach based on the sacred text and the authority of the ancients (Aristotle and the Church Fathers). It would be quite improbable, moreover, that anyone would be threatened with burning simply for a mistake in calculation.

We have to recognize, however, that the prohibition on Galileo's teaching did not prevent a certain freedom being exercised in the very Catholic kingdom of France. Descartes prudently went into exile, but Père Marin Mersenne and Pierre Gassendi each propagated in their way the accursed doctrine. The scandal provoked by Darwin's work was more profound.[86]

To sum up: for Galileo, the physical arguments that convinced him of heliocentrism were, apart from the elegance of the Copernican model, his own discoveries: the phases of Venus and the moons of Jupiter. On Earth's diurnal rotation, he correctly rebutted the argument of 'common sense' (which maintained that, if Earth was turning, we should perceive this: stones, for example, would not fall vertically),[87] and therefore made this rotation possible. He then mistakenly put forward the argument of the tides, and also, without convincing proof, the correct argument of the trade winds.

Subsequently, a whole raft of experimental demonstrations would confirm Galileo. On heliocentrism:

- the aberration of light, which was responsible for works dealing with heliocentrism being removed from the church's Index of prohibited books;
- the parallax of the stars;
- the Doppler-Fizeau effect.

On Earth's diurnal rotation:

- Jean Richter's measuring in 1679 of the variation of the period of a pendulum along a meridian;
- the movement of Foucault's pendulum in 1850;
- Reich's experiment of 1831.[88]

These direct proofs were the result of a wonderful chain of scientific advances that we shall now briefly describe.

Proofs of Earth's Motion

We shall first discuss the movement of Earth around the Sun, then its own rotation.

Around the Sun

The aberration of the stars

The aberration of the stars (or of light) is the effect of the finite speed of light and the relative motion of Earth in relation to the stars, this being Earth's own movement if we assume the stars to be fixed. Historically, the merit of having shown this effect (in 1727) is that of James Bradley (1693–1762), a British astronomer. This proof of Earth's orbiting played a historic role: as we noted, it convinced Benedict XIV in 1741 to give the *imprimatur* to the first edition of Galileo's complete works.[89]

It is sometimes very interesting to discover something different from what one was looking for. In this case, by looking for parallax, what was found was the aberration of light. The history of its discovery demonstrates this point. From the early eighteenth century it was established as scientific doctrine that Earth turns around the Sun, but scientists were still unable to refute by observation the major objection to this: if Earth moved in space, we would not see the stars (especially, those at the zenith) at the same angle six months apart, when Earth is at two opposite positions. There should therefore be a phenomenon of annual parallax. The attempt was made to demonstrate this in relation to a star assumed to be close. The Irishman Samuel Molyneux (1689–1728) chose γ *Draconis*, the brightest star in the Drago constellation. He did indeed find a variation in its position in the sky, but far greater than was expected, and in fact in the opposite direction from that of a parallax effect. After having eliminated all causes of possible error, Bradley proposed the following idea: since light takes time to spread, we only see stars after a certain interval; in other words, their apparent position does not indicate their real position.[90]

The classic image that helps to illustrate the aberration of the stars (or more precisely that of light) is the following: when rain falls vertically on a moving car, the driver sees it falling at an angle. Though the trajectory of the raindrops is perpendicular to the trajectory of the car (making an angle θ with the road), these drops are seen from the moving car as coming from the front, i.e., at an angle θ' (the apparent angle), which decreases with the car's speed. This difference between the real angle and the apparent angle is proportionate to the speed v of the car. The existence of this difference demonstrates the existence of this movement and even makes it possible to calculate its speed. Let us now replace the rain by the light rays emitted by a star, and the moving car by the moving Earth. How do we know the difference between the real and apparent angles of a star when we only have access to the apparent angle θ'? By waiting six months: if Earth is turning around the Sun, its speed v changes to $-v$ and the apparent angle becomes $-\theta'$. The difference between these two angles gives twice the difference between the real angle and the apparent angle (see calculation in Appendix A).

The parallax of the stars

One of the most convincing arguments against Earth's moving around the Sun – outside of Holy Scripture – was the absence of parallax. Galileo tried to measure this. All the stars, he believed, had the same intrinsic luminosity, and it was their distances from Earth that explained their apparent brightness. The brightest stars were therefore the closest ones, and consequently those most sensitive to the parallax effect, while the least bright stars could be used as a frame of reference. He failed in this attempt, and we now know that the distance of the stars is such that, with the base used by Galileo (the diameter of Earth's orbit), it would need a precision unattainable at this time to make a usable measurement.

The first measurement of the parallax of a star was achieved by Friedrich Wilhelm Bessel (1784–1846), who managed the tour de force of measuring in 1838 an angular difference of 0.3 seconds of arc.[91] He accordingly estimated the distance of star 61 in the constellation of Cygnus at 10.4 light-years (today's figure is 11.2). This was perhaps the finest measurement in the nineteenth century, overcoming an objection that was two thousand years old. (See calculation in Appendix A.)

The Doppler-Fizeau effect and the light of the stars

If the frequency of a sound wave is what characterizes sounds as high or low, the frequency of a light wave determines its colour (blue being a higher frequency, red a lower one). When the receiver moves in relation to the source, the frequency of the wave it perceives (apparent frequency) varies: if the receiver moves towards the source, this increases (sound is higher, colours become bluer); if it moves away, it decreases (sound is lower, colours redder). When this effect is produced with sound (for example, the siren of an ambulance approaching or receding), this is called the Doppler effect.[92] When it is produced with light (for example, the red shift of the galaxies that indicates the expansion of the universe), it is called the Doppler-Fizeau effect.[93]

To give an intuitive understanding of the Doppler effect, assume we have a swimmer moving away from the shore, so that the waves seem to reach him more quickly. More precisely, their apparent frequency (apparent, since the frequency with which they break on the beach remains constant) increases. As he returns to the shore, he moves in a sense with the waves, so that their apparent frequency is reduced. If we assume that there is in the sea a fixed source of waves (the action of the wind against the current on the water surface), it is the speed of the swimmer in relation to this source[94] that determines the change of frequency. Figure 35 shows two positions of Earth at a six-month interval:

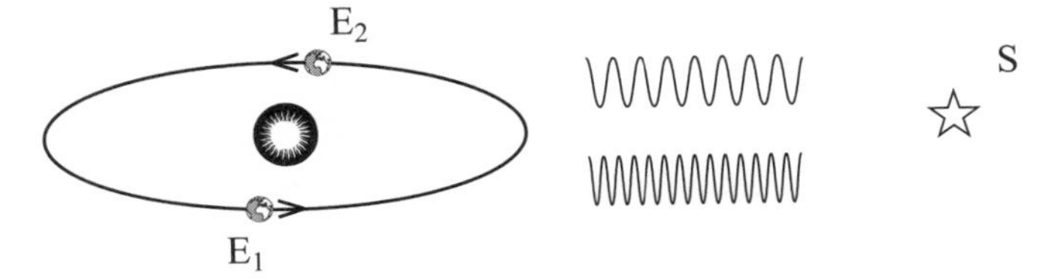

Fig 35. Doppler-Fizeau effect. When Earth moves away from star S, as at E_2, it receives from star S light of lower frequency than when it is at E_1, at which point it moves in a sense towards the light source.

S is a star assumed to be fixed (or in any case in constant motion in this interval of time), which for convenience is situated on the plane of the ecliptic. This star emits light rays characteristic of the chemical elements that compose it. Their frequency is perfectly known; for example, sodium has two distinct rays with wavelengths 589 and 589.6 nanometres. When Earth is at E_1 and moving towards the star, the apparent frequency is greater; conversely, it is lower and thus redder when Earth is at E_2. This shift in frequencies is a strong indication of Earth's orbiting (see calculation in Appendix A).

Earth's rotation on itself

Centrifugal force due to the diurnal rotation

A very simple element of proof for the existence of a centrifugal force, and thus of Earth's rotation, is the observation that planets that rotate on themselves are not exactly spherical, but slightly flattened at the poles. Is this the case with Earth? No for Cassini, who trusted Descartes's theory of vortices and the first geodesic measurements. Yes for Newton, who gave a theoretical prediction, seeing this distortion as created by centrifugal force at the time when the primitive Earth was still malleable. With a simple model, but using his theory of gravitation, Newton announced a flattening of 1/231, which new measurements on the ground confirmed.[95] This flattening of Earth is responsible for a reduction in gravity, *g*, of 0.20 per cent between the poles and the equator.

But the solidified Earth continues to turn, and so centrifugal force has the effect of further diminishing gravity *g* at the equator by 0.33 per cent. As we know that the period of a pendulum is inversely proportional to the square root of *g*, we can appreciate that these two reasons both prolong this period as the equator is approached. Jean Richter (1630–96) observed that a pendulum set to a period of one second in Paris slowed down by 2½ minutes per day in Cayenne, which was in agreement with the theory.

The inertia principle

For the Aristotelian physics of motion, which was more or less that of common sense codified and generalized, any body ends up stopping if no force is exerted on it. This is what experience seems to confirm. No matter how many times billiard balls are rolled on a horizontal surface, what materials the balls and the surface are composed of, or their initial speed, they still end up finally stopping. Galileo had understood in his own way that there is always a force of friction that cannot be totally suppressed. If this did not exist (a thought experiment), bodies in free motion would continue their course indefinitely in a straight line.[96] Newton subsequently explained that *only an applied force* can change velocity, either in magnitude or in direction. This is what in modern terms is called the principle of inertia. A ball released from the top of the mast of a ship moving uniformly in a straight line will continue, because of its initial speed, to accompany the ship with the same horizontal speed. That is why it still touches down at the foot of the mast, and not farther back. If we replace the ship with Earth, we understand why, according to Galileo, a stone thrown vertically into the air must fall back to its starting point.

The relativity of motion

How should motion be defined? It is characterized by both a direction and a speed, which only have their meaning in relation to a frame of reference. If this is changed, then the trajectory is changed, as we have seen in comparing the trajectories of the planets in the Ptolemaic and Copernican systems.

Seen from a moving bicycle, for example, the valve of the tire clearly describes a circle centred on the wheel hub. In relation to a frame of reference on the road, however, it follows a more complicated curve, that of a cycloid.[97] If we now consider a nail extruding from the tyre – clearly a thought experiment, as a real nail couldn't penetrate the road – its trajectory, seen from the bicycle, is again a circle, but with a radius slightly greater than the wheel. In relation to the road, however, this is a curve more complicated than the cycloid, that of the *trochoid* drawn in Figure 37. If the road was then a circle instead of a straight line, the previous trajectory would become a curve known as an *epitrochoid* (Figure 38; see Appendix E for the equations of these curves).

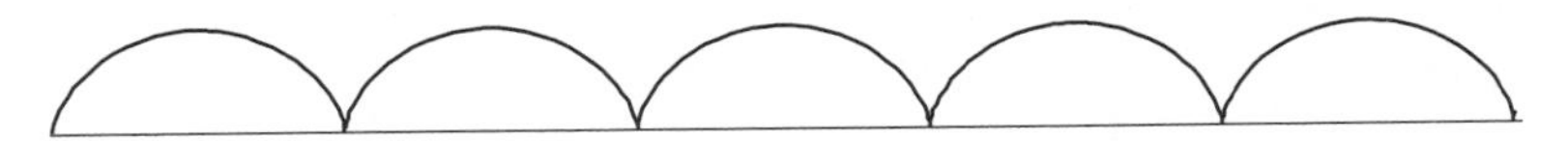

Fig. 36. Cycloid: the trajectory of the valve as seen from the road

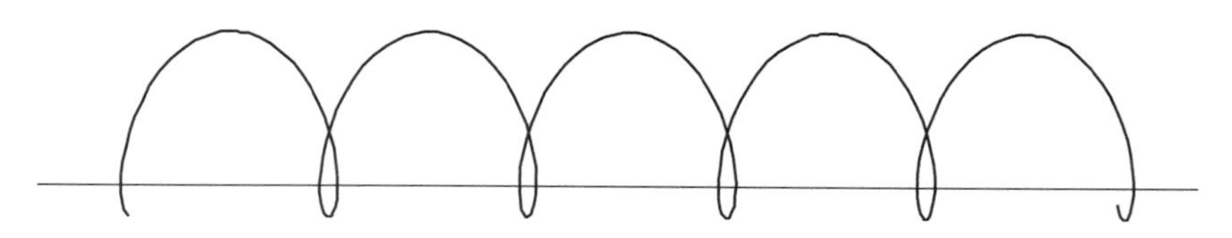

Fig. 37. Trochoid: the trajectory of the nail as seen from the road

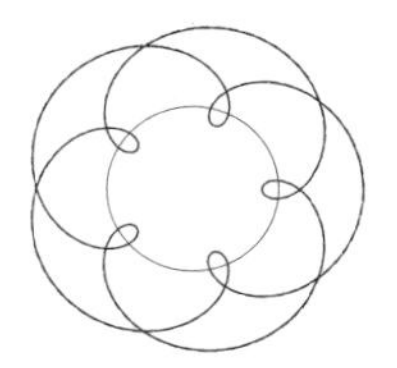

Fig. 38. Epitrochoid: compare with Fig. 45

The planets, then, describe *grosso modo* circles in relation to a frame of reference fixed on the Sun, but extremely complicated trajectories as seen from Earth (see Figure 45). There is more than an analogy here with the previous example of the cycloids.

By changing the frame of reference, velocity is also changed. We know today that the principle of inertia (see p. 140–1) is valid only for certain frames of reference known as Galilean. We also know that any frame of reference with a uniform motion in relation to such a Galilean frame of reference is itself Galilean. And we know, finally, that a system of axes centred on the Sun and passing through certain stars is a very good Galilean system (the principle of inertia is very well verified here). But what of a frame of reference fixed on Earth? Earth does not follow a uniform motion in relation to the Galilean frame of reference centred on the Sun, but a movement of revolution around the Sun and a movement of rotation on itself. If the frame of reference attached to Earth turns with this only in its movement around the Sun (which is then called a geocentric frame of reference), but without following its diurnal motion, this is still 'fairly' Galilean (the elliptical orbit of Earth is equated with a straight line traversed at a uniform speed). It is clearly less so if it also turns with Earth around its axis. Strictly speaking, then, the principle of inertia is not applicable to a movement established in relation to axes fixed on Earth and turning with it. As a consequence, a body in free fall (with no force exerted on it apart from gravity) does not exactly describe a straight line in a frame of reference that moves with Earth. The deviation is actually very tiny for small trajectories, and Galileo was more or less right. A fairly long fall is needed (Reich's experiment of 1830 was with a free fall in a pit of 158 metres), or the repetition of small movements (Foucault's pendulum experiment in 1851), to display this deviation.

One way of applying the principle of inertia, however, in non-Galilean frames of reference, is to introduce additional, fictitious drag forces, responsible for these changes of trajectory. Such are centrifugal force and the Coriolis force. The former is well known, what we feel in a car taking a rapid turn, or on a fairground ride. The Coriolis force is more subtle, only bearing on a body in motion in a particular (non-Galilean) frame of reference. It applies therefore to a body in free fall on Earth, giving this an eastward shift (for example of 2.58 centimetres in Reich's experiment).

Reich's experiment

This experiment[98] helps the intuitive understanding of the Coriolis force. It consists in verifying that a body in free fall does not drop quite vertically.

The circumference of a circle of radius r equals 2πr. So, if Earth made one turn per second, a point on the equator would travel a distance of 2πr in a second. Since we know that it actually makes one turn in 24 hours, this point travels a distance of $\frac{2\pi r}{(24\times 3600)}$ per second. This is its speed of rotation. Let us consider a pit of depth *h*. The bottom of this pit, being at the reduced distance r – *h* from the centre of Earth, thus travels at a speed of $\frac{2\pi(r-h)}{(24\times 3600)}$ per second, i.e., slightly less. The principle of inertia, applied to the geocentric frame of reference, stipulates that, in the absence of a horizontal force, the released ball has to continue with the same horizontal component of its velocity. But the base of the pit, still in this geocentric frame of reference, will have moved less quickly, and this is why the ball does not fall quite vertically, but with a slight easterly shift. See Appendix A for the calculation.

Foucault's pendulum

This historical experiment was conducted at the Panthéon in 1851, with a pendulum of length l = 67 metres and a period of approximately 16 seconds. It was observed that the plane of oscillation of the pendulum turned around a vertical axis with a period T. It is harder to make the effect of the Coriolis force in this experiment intuitive.

The principle of inertia stipulates that, in relation to a geocentric (thus quasi-Galilean) frame of reference, the plane of oscillations is fixed. To simplify, let us place the Panthéon at the North Pole. The frame of reference attached to the rotating Earth, *in relation to which these observations are made*, itself makes one turn in 24 hours. It will seem therefore that it is the plane of oscillation that turns: in fact we visualize this in a turning frame of reference that seems fixed to us as we are attached to it.

The full analytical calculation is complex, but the essential part of it will be understood from the simplified calculation given in Appendix A. We see that, in contrast to Reich's experiment, the effect is greatest at the poles, where the period is that of the rotation of Earth, T = 24 hours, and zero at the equator. Not only is this a laboratory experiment that proves the rotation of Earth, it also enables latitude θ of the experiment site to be calculated.

Proofs 'Beyond Debate'?

- The aberration of the stars *could* be caused by an identical movement of all the stars perpendicular to the ecliptic and with a period of a year;
- parallax could be due to a shift in the stars with a period of a year, this shift being proportionately greater for those stars close to Earth;
- the Doppler-Fizeau effect could be created by a periodic movement of the stars, particularly those situated on the ecliptic.

These mysterious annual movements of *all* the stars would still need explanation.

As for the proofs of Earth's rotation on itself, as distinct from the proofs of its rotation around the Sun, these are not geometric but dynamic. As we have seen, they call on the *apparent* violation of the principle of inertia due to the fact that the terrestrial frame of reference is not Galilean. This principle, too – however well established it is – could be false, and by developing flights of imagination, one could perhaps explain the experiments of Foucault and Reich in some ad hoc fashion, even by a variation in the period of the pendulum between the pole and the equator.

But the fact that several distinct phenomena can be related to a single cause constitutes a strong argument for taking this cause seriously and exploring whether other phenomena still unknown would not also follow from it.

In a very fine text published in 1905, Poincaré speaks of truth:

> '... Therefore,' have I said in *Science and Hypothesis*, 'this affirmation, the earth turns round, has no meaning ... or rather these two propositions, the earth turns round, and, it is more convenient to suppose that the earth turns round, have one and the same meaning.'
>
> These words have given rise to the strangest interpretations. Some have thought they saw in them the rehabilitation of Ptolemy's system, and perhaps the justification of Galileo's condemnation.
>
> Those who had read attentively the whole volume could not, however, delude themselves. This truth, the earth turns round, was put on the same footing as Euclid's postulate, for example. Was that to reject it? But better; in the same language it may very well be said: These two propositions, the external world exists, or, it is more convenient to suppose that it exists, have one and the same meaning. So the hypothesis of the rotation of the earth would have the same degree of certitude as the very existence of external objects.
>
> But after what we have just explained in the fourth part, we may go farther. A physical theory, we have said, is by so much the more true, as it puts in evidence more true relations. In the light of this new principle, let us examine the question which occupies us.
>
> No, there is no absolute space;[99] these two contradictory propositions: 'The earth turns round' and 'The earth does not turn round' are, therefore, neither of them more true than the other. To affirm one while denying the other, *in the kinematic sense*, would be to admit the existence of absolute space.
>
> But if the one reveals true relations that the other hides from us, we can nevertheless regard it as physically more true than the other, since it has a richer content. Now in this regard no doubt is possible.
>
> Behold the apparent diurnal motion of the stars, and the diurnal motion of the other heavenly bodies, and besides, the flattening of the earth, the rotation of Foucault's pendulum, the gyration of cyclones, the trade-winds, what not else? For the Ptolemaist all these phenomena have no bond between them; for the Copernican they are produced by the one same cause. In saying, the

earth turns round, I affirm that all these phenomena have an intimate relation, and *that is true*, and that remains true, although there is not and can not be absolute space.

So much for the rotation of the earth upon itself; what shall we say of its revolution around the sun? Here again, we have three phenomena which for the Ptolemaist are absolutely independent and which for the Copernican are referred back to the same origin; they are the apparent displacements of the planets on the celestial sphere, the aberration of the fixed stars, the parallax of these same stars. Is it by chance that all the planets admit an inequality[100] whose period is a year, and that this period is precisely equal to that of aberration, precisely equal besides to that of parallax? To adopt Ptolemy's system is to answer, yes; to adopt that of Copernicus is to answer, no; this is to affirm that there is a bond between the three phenomena and that also is true although there is no absolute space.

In Ptolemy's system, the motions of the heavenly bodies can not be explained by the action of central forces, celestial mechanics is impossible. The intimate relations that celestial mechanics reveals to us between all the celestial phenomena are true relations; to affirm the immobility of the earth would be to deny these relations, that would be to fool ourselves.

The truth for which Galileo suffered remains, therefore, the truth, although it had not altogether the same meaning as for the vulgar, and its true meaning is much more subtle, more profound and more rich.[101]

Part Three

'ONLY' SCIENTIFIC TRUTH?

This book has three particular aims, which are not independent of one another:

- to contribute to bringing scientific culture into general culture, and vice versa;
- to show how and why the scientists of the Renaissance, despite being good Christians, were forced to abandon the literalist reading of sacred texts;
- to rehabilitate the supposedly naïve notion of *scientific truth*, against the idea that science is no more than a socially constructed doxa.

The development of ideas about Earth has gone together with the history of thought as a whole, that of the 'humanities' as well as that of the so-called hard sciences. As an illustration of scientific procedure, the choice of studying Earth might seem a bad example. It certainly allows observations to be made, but can anyone produce an eclipse or a retrogradation in the orbit of Mars, let alone planetary creation? We are a good way from the conventional idea that science can only be built up by starting from an experiment that has to be repeatable to be credited as science, in order then to proceed to theory: a theory that, in its turn, must be refined in the light of the results of new experiments. The famous episode of a falling apple (they had been falling ever since Adam and Eve) no more enabled Newton to formulate the law of gravitation than new observations set Copernicus on the road to heliocentrism. A better image would be that of a loop with no privileged starting point, in which theories and empirical findings follow one another in turn. It is customary to say that, in the last analysis, experiment both decides and serves as motor. But however reasonable this sounds, the formula 'in the last analysis' can hide a multitude of subtleties. Copernicus' model did not account for quantitative observations any better than that of

Ptolemy. It did however make it possible to resolve some recurring problems in ancient astronomy, such as that of the phases of Venus, and above all it opened the way for Kepler and Newton. The superiority of the Copernican model over that of Ptolemy is thus not reducible to a mere competition over figures.

Despite being inaccessible to experiment, investigation of Earth's age and its motion pointed to problems that were eventually posed correctly, and found a scientific response resulting from the rational combination of numerous observations, experimental results and physical laws. This is the situation with the majority of present-day scientific questions. Few are completely experimental: every experiment or observation rests in its turn on a structured set of other experiments and supposedly *true* theories that they put to the test.

CHAPTER 8

Why Truth Matters

Guilty of the Word 'True'?

Do we then have any right to speak of *scientific truth*? The use of this expression may be shocking: it presupposes a notion of absolute that is evidently naïve.[1] The logic of this critique is that the term 'truth' should *never* be used outside of mathematics, since every truth has the propensity to evolve. I believe, however, that it is necessary at any particular moment to consider a proposition or property as either true, false, not demonstrated or even not demonstrable,[2] naturally on the condition of being ready to change one's view. We should note that this attitude is widely shared in practice by the human community (both scientists and others). Should the use of the adjective 'true' really be banned from laboratory life, or even from life in general? Paradoxically, it is this reference to the true, and thus to the false, that makes it possible to develop the content, meaning and limitations of truths, rather than the approach that would simply move through an erratic world[3] in which every statement is equally valid.[4] Truths, at least the truths of science, always have a domain of applicability, and progress has often consisted in displaying this. Newton's mechanics, for example, is true in a broad domain: for more than two centuries it was sufficient to successfully guide human activity, and continues to do so today. It ceases to be legitimate for velocities that approach the speed of light, or when dimensions or temperatures are too small. Recourse is then needed to relativity or quantum mechanics. In the same fashion, the determinism of Poincaré set practical limits to that of Laplace.

Scientists engaged in research have sometimes been called 'workers on proof'. But they are not alone in this. Can scientific truth be compared with judicial truth? In both cases, what is involved are assertions very generally accepted by the community in question, arising from procedures that are well codified and controlled. In justice as in science, procedures of deduction were used in the past that are now considered primitive: ordeal, the judgement of God, judicial duel. Eventually, the time of scientific rationalism arrived: first of all in the unfortunate form of phrenology,[5] then with fingerprints and today with DNA samples. And as in science, justice can fail to reach a conclusion or reach a wrong one; judicial truth may not coincide with truth in general, if it is powerless or goes astray. But no approach would be possible at all without the understanding that truth exists; that is what judges have to seek, and it is only in relation to this that it is possible to say whether or not they are mistaken. In this quest for truth, the judge, like the scientist, is helped by combining several different approaches.[6] But the comparison stops there: justice examines the cause (the responsibility) of human actions and accordingly determines a penalty as a function of culturally constructed rules. And so the force of prejudices, traditions and social interests has a decisive weight here.

The Conservatism of Science

Conservatism also plays a part in the sciences, but less than people sometimes say. They are not (or rather, they are no longer, since the end of Aristotle's sway) protected by the argument from authority. Every misunderstood scientist or para-scientist invokes Galileo[7] (more rarely, Einstein) as example of the scientific genius who was ignored by his fellow citizens and rediscovered long after. This is quite simply false. Galileo was seen in his time as a great mathematician, and his fame stretched beyond the borders of his country. A significant proportion of contemporary intellectuals (including clergy) were convinced that his views were correct. Unrecognized scientists discovered post mortem are quite the exception, such as the mathematician Évariste Galois, who died in 1832 after a duel caused by 'a well-known coquette',[8] or the father of modern genetics, the botanist monk Mendel (1822–84), who worked in isolation in Brno.

Taking account of the boldness of new hypotheses, such as the principle of inertia, the heliocentric system, universal attraction, quantum mechanics, relativity, etc., it is no exaggeration to maintain that these revolutionary theories were generally adopted in quite a short space of time. But science does suffer all the same from a certain inertia: no one is keen to abandon a construction that has cost so much to build up. New theories have to offer proof.[9] A certain resistance is inevitable and even healthy – as long as it is not obstinate. Concerning Earth's age, it was necessary to overcome the paralysing authority of Holy Writ, then that of Kelvin. But even this did not take all that long: the first was 'beyond discussion', and the other 'only' scientific. This conservatism equally restrains discoverers: three scientific works that 'shook the world' – Copernicus' *De revolutionibus orbium coelestium*, Newton's *Principia Mathematica* and Darwin's *On the Origin of Species by Means of Natural Selection* – were each published only twenty years after their conception, under the friendly or competitive pressure of Rheticus, Halley and Wallace, respectively. Was this for fear of their consequences?

The undeniable conservatism of science (often referred to for the occasion as 'official science') is relative: it does not prevent it from accepting such disturbing assertions as that 'it is just as if an electron can go through two holes at once', or the equivalence of mass and energy, Langevin's twins paradox,[10] etc.

If science does not accept the transmission of thought, the memory of water,[11] or cold fusion,[12] this is not because of its innate conservatism, nor because it fails to understand the mechanism – a very general situation, even for scientists: the curative properties of aspirin, for example, were understood well before their mechanism was understood. If science rejects these phenomena, therefore, it is because of the lack of proof of their very existence.[13] Scientists (at least good ones) certainly do not lack imagination. Feynman observed that in their description of the vacuum, physicists introduced immense quantities of electromagnetic waves propagated in it – not to speak of oscillations of the vacuum with their corresponding trail of creation and annihilation of particles. They showed more creativity than those who see an angel flying, when this is no more than a chubby little boy with two wings, representable without too much imagination.

Until the invention of the steam engine, people always understood the tools they were using. Electricity changed this situation completely. Most people who use television are completely incapable of explaining the principle involved even in the crudest way. If the conjuror's tricks amaze people, it is because the spectacle is unusual; yet the everyday operation of the mobile phone or the microwave oven is just as hermetic. If these miracles are accepted, why not others, for example, telepathy or astrology? Paradoxically, credulity lies in 'crediting' scientific advances.

Finally, science does not rest content with replying to questions. It also opens up areas of unsuspected ignorance: nothing is less evident, for example, than that 90 per cent of the universe is (perhaps) made up of matter and energy of an unknown kind (so-called dark matter and energy).

From Good Sense to Good Science

The human brain was shaped by natural selection, enabling man to deal with the immediate external world. The only magnitudes it is able to imagine a priori, therefore, are those on the human scale,[14] whether we are talking of distances, speeds, masses or durations. We can imagine 1 metre, 1 kilometre per hour, 1 kilogram, 1 second, or even a million times more or a million times less: but scarcely any further. The same holds for quantities without dimension: who can represent to themselves the billions upon billions of stars in the universe, or, perhaps still worse, the Avogadro number (6.02×10^{23}), which is the number of atoms contained in 12 gms of carbon?

The ancients possessed a very anthropological collegium of gods and goddesses, and were used to a physics that was not only on the human scale but also had human connotations: the planets, for example, were endowed with qualities, even with soul. To use Pierre Duhem's expression, this was 'a physics that started from common sense to end up with common sense', the physics of Aristotle being its most complete expression, at least as regards falling bodies. But dimensions below a billionth of a metre require a new physics: quantum mechanics, which is completely removed from common sense. Tremendous speeds and colossal masses necessitate the aid of relativity.[15]

Long timeframes did not actually give rise to a new physics, but their integration into geology was very difficult. If Buffon never published his estimates of Earth's age in terms of millions of years, this was not so much from fear of the ire of the church, as of the immensity of the figure. The integration of long times into thought made it possible to understand how events that are extremely rare on a human timescale become almost certain in the long term, and can then constitute an evolutionary motor.

Heliocentrism and the theory of evolution were not combatted simply in the name of dogma, but also of good sense, almost in the literal meaning: what the senses lead us to understand.

The evolution of our ideas about Earth reflects very well the advances of science: it is a question of understanding changes of scale.[16] The initial conceptions were locally satisfactory and independent of time, for example, the idea of a flat Earth without development, at the centre of a small universe; it followed as a law of gravity that *the natural tendency of a heavy body is to descend to the place that is proper to it* (Aristotle). They would find their accomplishment (not, of course, their achievement) in the conceptions of today.

This primitive view of nature on the human scale is matched by an equally pre-scientific explanation: the order of nature, and its complexity, results, like that of society, from the will of its leaders, or even a great planner.[17] As far as the Christian world is concerned, the long quotation from Rev. Paley (note 51, p. 234) is characteristic of this view, which is still today, with an ad hoc wrapper, that of the champions of 'intelligent design'.

Before Darwin, it was not understood that the enormous complexity of the living world was the result of a *very long process*: natural selection. What made it possible for the theory of evolution to develop was taking into account the very long timeframe of physics. Just think of everything that a simple cell 'knows how to do', incomparably more complicated than a clock. That is why Pasteur, who did not refer to Darwin, liked to say that 'a little science estranges people from God, much science leads them back to him'.

By a totally different mechanism,[18] the 'marvellous order of the planets' – which for Newton could only arise from the will of the Lord – was also the result of a very long process, in the course of which a cloud of dust and gas evolved into our very organized solar system. The mechanism that creates order is now rather well understood;[19] it implies, among other things, the theory of deterministic chaos. Digital simulation made possible by powerful computers has made a great contribution to this: processes that took millions or even billions of years are represented in a few minutes or days.

Finding What You Are Not Looking For

This history of ideas about Earth illustrates the famous aphorism that the laser was not invented by trying to perfect the candle – or how questions that seemed insoluble find answers in a totally different context, and even how answers arise to questions that were not posed. It is commonplace to seek for what has not been found, but the salt of research is to find what you were not looking for.[20] Thus it was held to be:

- impossible to determine an absolute age for Earth

(but the discovery of radioactivity in a domain that was a priori quite foreign to this changed the terms);

- impossible to conceive of and therefore to determine an absolute motion for Earth

(but discovery of the Coriolis force showed the meaning that could be given to this);

- out of reach to know the chemical composition of the stars[21]

(but spectroscopy brought the answer).

The fruitfulness of encounters between different fields – palaeontology, physics, history, astronomy, etc. – no longer needs demonstrating:

- considerations on the development of living species led to doubt being cast on Earth's age, which had seemingly been well established by physicists: that was Darwin's victory over Kelvin;
- study of striations on shells made it possible to determine the length of the lunar cycle in a very remote past, and thereby gave an estimate of the distance of the Moon from Earth several hundred million years ago;
- purely astronomic considerations such as the precession of the equinoxes make it possible to date historical events.

A political consequence follows. It is dangerous to believe – or have others believe – that a more utilitarian public policy would make basic research in science more useful to society. The risk on the contrary is of making it sterile.

Scientists Don't Always Do Science

The purpose of this book was not to write a history of science. Deliberately limited to the logic of the development of ideas about Earth, it has not examined society and history, something that historians of science certainly do not neglect to do.[22] Blamont, in his monumental work *Le Chiffre et le songe*,[23] even writes that the Galileo affair, generally presented as a symbol of the confrontation between religious obscurantism and science, was above all a political quarrel 'between the house of France and the Habsburgs, between Gustav Adolf and Wallenstein, between Jesuits and Dominicans, and above all between two old men [Urban VIII and Galileo] who were each cursed with an overweening pride'. Maybe Blamont is bending the stick rather in the other direction.[24]

However, by glossing over the social context in the history of ideas, have we not somewhat idealized the construction of science, as an intellectual tournament far above the real world and insensitive to social determinations?

In fact, as Alan Sokal has put it very well,[25] there is a confusion in the use of the word 'science'. This can denote, among other things:

1. Beside the sum of acquired knowledge, the rational search for laws making it possible to understand (and act on) the processes of nature (or even of society) and leading to universal results, i.e., independent of the personality of the individual announcing them (even if he is generally today male, white, writing in English and from a quite privileged social position); these laws are thus in principle testable by any fraction of the human community.
2. The public[26] and private institutions intended to organize and finance it (with the weight of corresponding social and political interests).
3. The set of practical results of research projects, stretching from the invention of the BCG vaccine to the neutron bomb or transgenic maize,[27] often referred to as technoscience.

These three meanings are connected: scientific research (1) is the work of men living in society and organized for the most part today into powerful bodies (2), who have to justify themselves socially (3). A relativism based on definitions (2) and (3) of science is quite justified: scientific research should not be confused with the ministry for scientific research (or worse still, with the minister for scientific research!). Would anyone identify justice with the ministry of justice? But relativism can lead to a sterile scepticism, even obscurantism, when applied to meaning (1) of science. Bertrand Russell would probably have classified science in sense (1) in the category of 'knowledge' and in senses (2) and (3) as 'power'. These categories are not watertight: power clearly has its effect on knowledge of various kinds. But their developments follow different or even opposing logics. We need only consider the status of free circulation of information and the question of patents. Certainly, definition (1) represents an ideal. Scientific research in the first sense is the domain of scholars who are men (more rarely, women),[28] sharing to a greater or lesser degree the prejudices[29] of their time. One example borders on caricature: when the great Cuvier spoke of blacks as 'the most degraded of human beings' (note 45, p. 215), what he expressed were the caste prejudices of a peer of the realm under Louis-Philippe; relativists conclude from such things that 'you can make science say anything'. We believe it is more useful to make the effort to *demonstrate* that in statements such as this Cuvier purely and simply abandoned the terrain of science in sense (1) above. Cuvier had gone off the rails. Although to speak of derailment, you need first of all to define the rails.

The definition of such rails is the task of the philosophy of science, a very wide field that the present book will not tackle,[30] even if the subjects tackled have made certain incursions inevitable.

Despite caution being advised as to what history can 'teach',[31] it does have the virtue of casting light on aspects of (post-)modern thinking and the revival of religious fundamentalisms, which I see as reactions that are understandable but reactionary in every sense of the word – that is, with no future, or at least not one to be desired.

All (or almost all) scientists of the Renaissance were initially, we could say, creationists.

Just as anachronistically, we could also maintain that many scientists of the eighteenth and nineteenth century (including Newton and Kelvin) upheld a form of 'intelligent design'. The Darwinian revolution, by exploiting all the potentialities of *deep time*, marked the end of this belief, as far as scientists were concerned. The great watchmaker beloved of the Reverend Paley could very well be blind, as Richard Dawkins neatly put it.[32]

Understanding the difficult chain of reasoning that led to – one could almost say 'compelled' – present-day knowledge can be helpful in combating the retrograde movement promoted by creationism and the different forms of literalist reading of the sacred texts.

When it leads to a methodological equivalence between science and religion[33] (or even magic), scientific relativism opens a royal road to religious conservatisms: what weight do the objections brought up by scientists (referred to pejoratively) have, when science can be made to say anything? The history of the controversies about Earth refutes this aphorism, even if it happened at some moments that scientists did say all kinds of things (and science, in its applications, did all kinds of things).

An intellectual misery often fuelled by misery pure and simple, the resurgence of the various religious fundamentalisms makes Galileo's arguments and Darwin's contribution astonishingly current today.

Afterword

When Proust describes in *Swann's Way* the reaction aroused in M. Vinteuil by the conduct of his daughter – it would certainly have been called misconduct by conventional morality – he observes that the world of beliefs behaves, in relation to that of facts, rather as something separate and independent on which the latter exerts scarcely any action:

> And yet however much M. Vinteuil may have known of his daughter's conduct it did not follow that his adoration of her grew any less. The facts of life do not penetrate to the sphere in which our beliefs are cherished; they did not engender those beliefs, and they are powerless to destroy them; they can inflict on them continual blows of contradiction and disproof without weakening them; and an avalanche of miseries and maladies succeeding one another without interruption in the bosom of a family will not make it lose faith in either the clemency of its God or the capacity of its physician. But when M. Vinteuil thought of his daughter and himself from the point of view of society, from the point of view of their reputation, when he attempted to place himself by her side in the rank which they occupied in the general estimation of their neighbours, then he was bound to give judgment, to utter his own and her social condemnation in precisely the same terms as the most hostile inhabitant of Combray; he saw himself and his daughter in the lowest depths.[1]

We may note that in this passage, where Proust gives the impression of defending an argument that is both quite general and particularly radical about the way in which faith seems capable of leading its own life in a sphere into which facts do not enter, and where it is accordingly immunized against the threat that they could represent for it, he does not apparently question in any way the existence of a world of facts that also has its own life and that nothing authorizes us to suppose can be reduced to a mere reflection or projection of the beliefs that we entertain about it. We may certainly not conclude, from the observation that these facts do not penetrate into the world of beliefs, and are consequently equally incapable of either producing or destroying them, that beliefs possess, for their part, the power to penetrate in some way into the world of facts, and are capable both of making these exist and of abolishing certain of them.

Nothing in what Proust says, moreover, brings into question that even moral reality may contain facts that have a certain objectivity and are not simply the product of the beliefs that we form in this domain. It is completely possible, we are given to understand, that the reprobation which Mlle Vinteuil's behaviour arouses is essentially of a social nature, and does not contain anything very much that is genuinely moral, still less morally justified. But what interests the novelist is the fact that her father, who manages to preserve a completely idealized image of her, despite all that he probably knows, applies to her a moral judgement that seems to have no relationship to the facts in question, which should logically lead him also to condemn her, whereas the social judgement that he expresses, on the subject of the very degraded position that his daughter and himself have come to occupy from the point of view of society and on the scale of reputations, despite being no more objective, does at least preserve a certain relationship with reality, sinning only by excess, as it situates this position on a level that can only be, to his eyes, the lowest of all, that which they would actually occupy if all the inhabitants of Combray were as ill disposed towards them as the worst of these are.

For quite a good while now it has become almost a commonplace, even in epistemology, to say that we cannot hope to find in reality anything other than we ourselves bring to it, and that it is in the end far more configured, influenced, and perhaps even quite simply produced, by the beliefs that we entertain about it than the latter can be by the former. As we have seen, however, this is not what Proust says, since in the passage I quoted he continues to speak of facts and the independence that they enjoy in relation to the world of our beliefs, in a way that could well seem singularly naïve to the enlightened postmodernists we have supposedly now become.

To put it frankly, all the efforts of postmodern criticism seem to have essentially led, on this question, to a result far less revolutionary and certainly more modest than is imagined most of the time: in other words, that the beliefs we form about reality are not determined uniquely by reality, and that reality, in certain cases, only plays a very small part, sometimes quite simply no part at all, in what we believe. Presented in this form, however, the result would clearly not seem very impressive, and the upholders of postmodernism are consequently obliged to replace it by another contention, which would certainly be forceful enough if, contrary to the facts of the case, it actually had been established. As Paul Boghossian very rightly puts it:

> To concede that no one ever believes something solely because it's true is not to deny that anything is objectively true. Furthermore, the concession that no inquirer or inquiry is fully bias free doesn't entail that they can't be more or less bias free or that their biases can't be more or less damaging. To concede that the truth is never the only thing that someone is tracking isn't to deny that some people or methods are better than others at staying on its track.[2]

What makes things particularly depressing in situations of this kind, which are unfortunately all too current today in philosophy, including unfortunately in the philosophy of science, is precisely the fact that the striking novelty that is proclaimed remains essentially in the state of mere proclamation, and that what has managed to get beyond this step gives the vexatious impression of coming down in the end to something completely familiar and even, to be honest, quite trivial. This is the observation that Alan Sokal makes of a passage in *Science in Action* by Bruno Latour, in which the author develops seven rules of method for the sociology of sciences:

> Here is his Third Rule of Method: 'Since the settlement of a controversy is the *cause* of Nature's representation, and not the consequence, we can never use the outcome – Nature – to explain how and why a controversy has been settled.'[3]

Note how Latour slips, without comment or argument, from 'Nature's representation' in the first half of this sentence to 'Nature' *tout court* in the second half. If we were to read 'Nature's representation' in both halves, then we'd have the truism that scientists' representations of Nature (i.e., their theories) are arrived at by a social process and that the course and outcome of that social process can't be explained simply by its outcome. If, however, we take seriously 'Nature' in the second half, linked as it is to the word *outcome*, then we would have the claim that the external world is created by scientists' negotiations: a claim that is bizarre, to say the least, given that the external world has been around for about 10 billion years longer than the human race. Finally, if we take seriously 'Nature' in the second half but expunge the word *outcome* preceding it, then we would have either (1) the weak (and trivially true) claim that the course and outcome of a scientific controversy cannot be explained solely by the nature of the external world (obviously some social factors play a role, if only in determining which experiments are technologically feasible at a given time, not to mention other, more subtle social influences) or (2) the strong (and manifestly false) claim that the nature of the external world plays no role in constraining the course and outcome of a scientific controversy.[4]

Contrary to what one might be tempted to believe, the criticisms formulated by people such as Sokal and Boghossian have nothing in the way of caricature about them, or even mere exaggeration. They give an idea that is sadly quite faithful to the disturbing – to say the least – level that epistemological literature of postmodern inspiration generally occupies, in terms of conceptual exactness and rigour of argument. The 'slippages' that Sokal discusses are in no way regrettable accidents, such as occur from time to time in an understandable way and could very readily be excused; they seem rather to constitute an actual style of thinking, whose powerfully original and completely innovative character can clearly escape only reactionaries and philistines of the worst water.

Another example of these 'ordinary' slippages is that which occurs when Latour speaks, as he regularly does, of the transformation, successful or otherwise, of a sentence or proposition into a fact: forgetting, or making a semblance of forgetting, that this way of expressing himself is quite simply devoid of sense. Strictly speaking, a sentence or a proposition can end up being recognized as factual, as representing a fact, but it certainly cannot be transformed, by no matter what wave of a magic wand, into a fact – except on the understanding that one is ready to view the world or reality as itself made up once and for all of facts and propositions. One can say of a true proposition that it represents a fact, but certainly not that it is a fact, by surreptitiously claiming the right to identify it purely and simply with the fact that it is deemed to represent.

It is difficult therefore not to leap out of one's chair on reading assertions like the following:

> A sentence may be made more of a fact or more of an artefact depending on how it is inserted into other sentences. *By itself a given sentence is neither a fact nor a fiction; it is made so by others, later on.*[5]

You may have written the definitive paper proving that the earth is hollow and that the moon is made of green cheese, but this paper will not become definitive if others do not take it up and use it as a matter of fact later on. You need *them* to make *your* paper a decisive one.[6]

> We need others to transform a claim into a matter of fact. The first and easiest way to find people who will immediately believe the statement, invest in the project, or buy the prototype is to tailor the object in such a way that it caters to these people's explicit interests.[7]

In the slippage we are dealing with here, from something that is a component of language (a statement) to a component of reality (a fact), the trivial starting point, which no one would think to challenge, is that it is impossible to refer to a fact except by way of a statement that describes it, and the switch consists in treating the statement as if what it states were equivalent to saying that instead of speaking of a fact, one may just as well speak of a factual statement, an assertion that, unfortunately, is by no means trivial and amounts in fact to a patent absurdity. In order for something to become considered as an established scientific fact, what is actually needed is that scientists manage to come to agreement on a positive response to the question of knowing whether it should or should not be recognized as a fact of this kind. But the power to decide what should be recognized as an established fact, at the outcome of a process that cannot in fact remain strictly individual but proves most of the time to be complicated and uncertain, has nothing in common with that of deciding what is a fact and what is not, something that, *pace* Latour, very much does depend on reality or nature, and is decided only by them. If the earth is not hollow and the moon not made of green cheese, it is not we who decided it this way, but reality, such as we believe we have come to know it. Unless, of course, we have decreed from the start that there can be no other use of the word *fact* than that in which a real or objective fact means a fact that we have recognized, or rather, decided, is one, in which case there is quite simply no sense in speaking of facts that are objectively realized or not, without being at the moment in a position to know which is the case.

To believe Latour, scientists who are involved in a controversy do not use nature as arbiter to try to settle this, which means that the philosophy, history and sociology of the sciences should also in principle not seek to do so. For the reasons indicated, it is an illusion to believe that using nature as external and impartial judge is precisely what the scientist is seeking to do in conducting experiments and testing hypotheses. One of the non-negligible secondary benefits of the operation is that it is possible in this way to transform oneself into a convinced realist, in the only genuinely understandable and acceptable sense of the term. As Latour puts it:

> We will then need to have two different discourses depending on whether we consider a settled or an unsettled part of technoscience. We too will be relativists in the latter case and realists in the former. When studying controversy – as we have so far – we cannot be *less* relativist than the very scientists and engineers we accompany; they do not *use* Nature as the external referee, and we have no reason to imagine that we are more clever than they are.[8]

Unfortunately, the writer of these lines does not seem to appreciate that decreeing that questions we have not yet managed to decide, and perhaps will never manage to decide, have not actually been decided and cannot have been decided in one sense or the other by reality itself, consists precisely in adopting a position that has nothing realist about it, and is even the very opposite of realism. For a genuine realist, what we are trying to know is precisely what the answer may be, independently of our cognitive activities, our interests and our preferences, to those questions to which we do not yet know the answer, but hope to find this one day. If the only part of science authorized by a realist interpretation was the stabilized part, this interpretation would precisely no longer be in any way realist, and the reality to which it refers, far from being capable of steady exploration and discovery, as realists who deserve the name believe, would rather give the impression of having come into the world as a function of the resolution of controversies, which would effectively explain why it is perfectly vain to count on it to help us to settle these.

Sokal gave as an epigraph to the article I have cited, on what one can and cannot conclude from the amazing facility with which he succeeded in mystifying, by a memorable practical joke, the review *Social Text*, an extract from a book by Larry Laudan in which the author explains that his real target was

> those contemporaries who – in repeated acts of wish-fulfilment – have appropriated conclusions from the philosophy of science and put them to work in aid of a variety of social cum political causes for which those conclusions are ill adapted. Feminists, religious apologists (including 'creation scientists'), counterculturalists, neoconservatives, and a host of other curious fellow-travellers have claimed to find crucial grist for their mills in, for instance, the avowed incommensurability and underdetermination of scientific theories. The displacement of the idea that facts and evidence matter by the idea that everything boils down to subjective interests and perspectives is – second only to American political campaigns – the most prominent and pernicious manifestation of anti-intellectualism in our time.[9]

A good part of what one can read in this field actually leads inevitably to wondering whether, instead of using arguments drawn apparently from the philosophy, history and sociology of the most serious sciences, but which one realizes almost at first glance show that these actually prove nothing of what is claimed, it would not have been more honest on the part of the authors, all things considered, to begin by simply explaining that their essential goal was to try and discredit, morally and politically, a certain number of traditional notions, such as for example those of truth and objectivity, which should be viewed not only as useless, but also as reactionary and pernicious. Latour has no hesitation in maintaining that

> Scientific or technical texts ... are not written differently by different breeds of writers [from those who write newspaper articles or novels]. When you read them, this does not mean that you quit rhetoric for the quieter realm of pure reason. It means that rhetoric has become heated enough or is still so active that many more resources have to be brought in to keep the debates going.[10]

But if this is really the way that things happen in the sciences themselves, we can clearly not expect the rhetorical means used, in their own rise to power, by those who seek to dispossess science of the prestige and influence that it has, so they say, largely usurped, to be more respectable and bring us closer towards the calm waters of pure reason.

It is rather the opposite that has every chance of coming about, and what we risk facing can scarcely be anything but additional confirmation of the fact that, even in those confrontations that are seemingly most intellectual, power is generally conquered not because one is capable of producing the best reasons, but rather because these reasons were the best for conquering it effectively, which the postmodern detractors of science seem for the moment to be largely doing, by methods that they would indeed say are probably, all things considered, neither better nor worse than those that generally permit a scientific theory to win out over its rivals.

According to Boghossian, the *succès de scandale* that Sokal's joke enjoyed shows three important things:

> First, that dubiously coherent relativistic views about the concepts of truth and evidence really have gained wide acceptance in the contemporary academy, just as it has often seemed. Second, that this onset of relativism has had precisely the sorts of pernicious consequence for standards of scholarship and intellectual responsibility that one would expect it to have. Finally, that neither of the preceding two claims need reflect a particular political point of view, least of all a conservative one.[11]

These three assertions, to my mind, are perfectly exact, and useful, I believe, for the reader to keep in mind. I consider as self-evident, in particular, that no more than the conceptions developed by postmodern authors in the field of epistemology and the philosophy of sciences actually imply the progressive political consequences that seem to them to result from these in a more or less automatic fashion, do the theses defended in the book imply in any way whatsoever the politically reactionary consequences that some will probably seek to draw from them, in the hope of managing in this way to discredit them. In both cases, unfortunately, what we are dealing with relates far more to the mere association of ideas consecrated by a habit rather too speedily acquired than to a genuine line of argument, with the result in the second case of culpability and unacceptability by association, and in the first of an innocence and legitimacy that are likewise by association.

Krivine's book tackles, taking a particularly well-chosen example and studying this in an extremely precise and detailed manner, the key question of which I have mentioned certain aspects:

> Are we entitled to say that Earth's age is 4.55 billion years, and its trajectory an ellipse centred on the Sun, with an average radius of 150 million kilometres? The majority of educated people today will say yes. Curiously, however, the fact that these assertions constitute what it is customary to call 'scientific truths'

is often perceived, three hundred years after the century of Enlightenment, as naïve or even improper. And it is actually *very* educated individuals who say this.[12]

Among the number of reasons that have led us to where we are today, there figures in key place, as the author notes, the fact that the attempt has often been made to select, and have taken for unchallengeable scientific truths, assertions that are certainly not scientific and certainly not beyond challenge. But despite the considerable popularity enjoyed by the argument, particularly among '*very* educated individuals', that concludes from this the need to seriously reconsider the importance that we have until now felt bound to extend to the notion of truth, it does not hold up very long to serious examination. For the conclusion we should rather feel obliged to draw is precisely that we need more than ever to preserve in all strictness, and reassert constantly, the distinction between what manages to win recognition and acceptance as truth at a given moment and in given conditions, without necessarily being so, and truth *tout court*. In other words, there really is no honest way of transforming the idea that a countless number of false things have managed to win temporary acceptance as true and even, in certain cases, as scientific, into an argument capable of being used against the idea of truth itself.

The title of this book already indicates clearly that its author is among those not resigned to seeing the constructions of science treated more or less as mere myths, whose superiority over others, just as plausible and respectable, in no way lies in their established truth, but simply in the privileged and even exclusive position they have managed to win and preserve in cultures of a certain kind. One of the main objectives of this book, accordingly, is 'to rehabilitate the supposedly naïve notion of *scientific truth*, against the idea that science is no more than a socially constructed doxa' (p. 147). Regarding the two examples that are treated here with impressive mastery and authority, the reader who might have doubted it will be convinced, I hope, that there can be and really has been, in certain cases, a gradual transition from myth to knowledge, or from mythical belief to scientific knowledge, leading to the displacement of the former by the latter, for reasons that are in no way arbitrary and do not simply bear on competition for power and influence between views that are intrinsically neither more nor less true than one another.

The thesis of a 'methodological equivalence' between science on the one hand, and religions and myths on the other, may rightly be considered as refuted in an exemplary way by the history told in the following pages:

> When it leads to a methodological equivalence between science and religion (or even magic), scientific relativism opens a royal road to religious conservatisms: what weight do the objections brought up by scientists (referred to pejoratively) have, when science can be made to say anything? The history of the controversies about Earth refutes this aphorism, even if it happened at some moments that scientists did say all kinds of things (and science, in its applications, did all kinds of things).[13]

As a number of its critics have noted, postmodernism is particularly marked by a certain tone and attitude that are already implicitly contained in the term *postmodern* itself, 'namely, the conviction that we are now too sophisticated to be moved by the intellectual ideals or accomplishments of the last three centuries'.[14] In particular, we are too educated and too sophisticated to be able to believe still in such things as reality and the progress of objective knowledge. Hubert Krivine holds this to be completely unfounded, and he is equally convinced that, if the science that some people with a certain contempt call 'orthodox' or 'official' remains, in essentials, faithful to the ideals and achievements that he refers to in the passage just quoted, this is certainly not out of any lack of imagination that prevents it from completely freeing itself from outdated constraints that the modern mentality for a long time imposed on it, and that we should today see ourselves as having gone beyond:

> If science does not accept the transmission of thought, the memory of water, or cold fusion, this is not because of its innate conservatism, nor because it fails to understand the mechanism – a very general situation, even for scientists … Scientists (at least good ones) certainly do not lack imagination.

Few absurdities seem to me to have known such great favour in recent times, and to be as capable of arousing such justified exasperation, as that which believes it is enough to free the imagination from every kind of obstacle in order to have a serious chance of immediately increasing considerably the number of scientific discoveries. It may well be, indeed, that some scientists lack imagination in certain circumstances. But it is certainly not only or even mainly because of a lack of imagination that the number of major discoveries may seem to be on the decline. It is perhaps also and even above all because reality is hard to know and, as is said, reveals to us its secrets only slowly and at the cost of considerable efforts.

But in order to be able to say this, it is of course necessary, as Krivine rightly does, to choose to continue expressing oneself resolutely in the way that the postmodern opponents of scientific realism consider irremediably 'naïve'. If what we call 'reality' is never more, in fact, than the result of an ever-shifting construction, the conception and realization of which is entirely our own responsibility, and if there really is no more difference than Latour suggests between what the inventors of fictional stories do and what the creators of scientific theories do, it is not very clear then in the name of what one should prevent oneself suggesting that scientists use their imagination with the same kind of freedom as novelists do, which would certainly open for science perspectives that are not only far wider, but also far more stirring.

Jacques Bouveresse
Paris, October 2010

APPENDIX A

The Proofs of the Earth's Motion

ABERRATION OF LIGHT

This aberration describes the fact that the apparent direction of a light ray coming from a distant star depends on the relative velocity of the observer.

Let R denote a frame of reference (Ox,Oy) containing a star S fixed in R. Let R' denote a frame of reference $(O'x',O'y')$, centered on the Earth, moving with uniform velocity[1] $\vec{v}$ along the Ox-axis. Let $\vec{c}$ and $\vec{c}'$ denote the velocity vectors of a light ray issued from E in the two frames of reference, and θ and θ' denote their polar angle respectively. The *classical* law of composition of velocity vectors:

$$\vec{c}' = \vec{c} - \vec{v}$$

upon projection on the axis of R', leads to:

$$c' \cos\theta' = c\cos\theta - v$$
$$c' \sin\theta' = c\sin\theta,$$

where θ' denotes the angle *as measured by an observer on the Earth.* Combining the above two equations leads to:

$$\tan\theta' - \tan\theta = \frac{v\tan\theta'}{c\cos\theta}.$$

Let us now calculate the difference $d\theta = \theta' - \theta$ under the assumption (realistic, as we shall see) that $d\theta \ll 1$. In the r.h.s. of the above equation, one can identify ϑ' with ϑ (to first order), which leads to:

$$\tan\theta' - \tan\theta = \frac{v\sin\theta}{c\cos^2\theta}.$$

Next, to first order in $d\theta$, one can write:

$$\tan\theta' - \tan\theta \simeq \frac{\theta' - \theta}{\cos^2\theta}.$$

Comparing the two expressions then leads to:

$$d\theta = \frac{v}{c}\sin\theta.$$

Remarks:

i) The aberration does not depend on the distance of the star, but on its direction only; in contrast to the Doppler-Fizeau effect, it is maximum when the direction of the star is vertical ($\theta = \pi/2$).

ii) The measurement of $2d\vartheta$ is done in a time interval of 6 months, when the velocity $\vec{v}$ changes into $-\vec{v}$.

iii) The calculation of the aberration requires the knowledge of the velocity $v = \omega R$ of Earth around the Sun. The pulsation ω was of course known: $\omega = \frac{2\pi}{365 \times 24 \times 3{,}600}$ rad/s, but the average distance Earth–Sun R was not. This explains the (relative) lack of precision of the velocity of light deduced from the calculations Rømer performed in 1676 (220,000 km/s instead of 300,000 km/s).

iv) The calculation presented here is non-relativistic, which is correct given the smallness of the ratio v/c. The first relativistic correction is of order v^2/c^2, hence negligible.

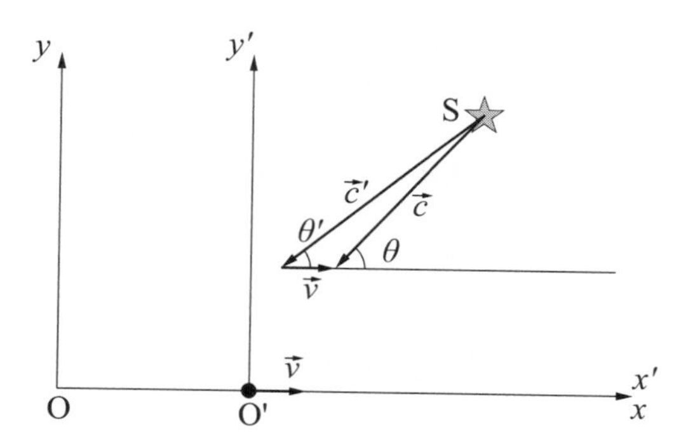

Fig. 39. *Earth in O' moves with a velocity $\vec{v}$. An observer on Earth will see the star S from an apparent angle $\theta' \neq \theta$.*

v) A daily aberration also exists, due to the rotation of Earth in 24 hours, but it is at least 60 times smaller (at the equator where it is maximum).

vi) The aberration constant v/c is of the order of 20" or arc; it is certainly small, however much larger than the 1″ parallax of the closest stars (see below). This explains why it has been identified well before the latter.

Parallax

The parallax is the angle under which two positions of Earth on its orbit six months apart are seen from a neighboring star.

Proxima Centauri is the closest star to the Sun. Its distance D to Earth has been obtained by triangulation, using twice the distance Earth–Sun as basis (see Figure 40). Its parallax has been determined to be $\vartheta = 0.772$ seconds of arc. The distance D is then given by:

$$D \simeq \frac{R}{\theta} = \frac{8}{60 \times 24 \times 365} \frac{180 \times 3600}{\pi \times 0.772} \simeq 4.1 \text{ light-years}$$

The distance Earth-Sun has been taken as 8 light-minutes, i.e., $\frac{8}{(60 \times 24 \times 365)}$ light-year, and the angle ϑ, expressed in radian, is equal to $\frac{\pi \times 0.772}{180} \times \frac{1}{3600}$.

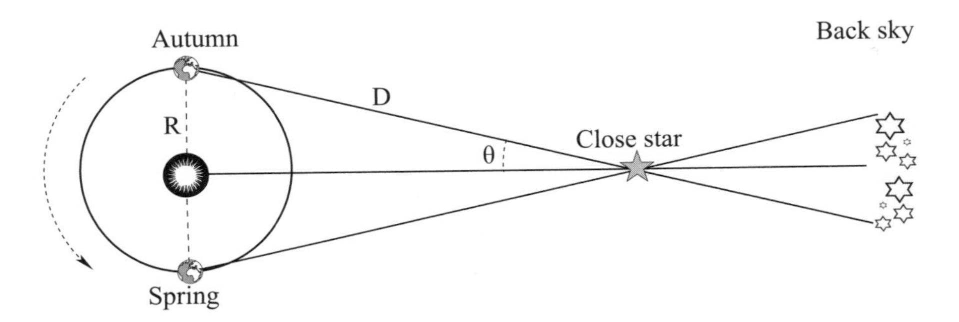

Fig. 40. *Principle of the method for measuring the distance of Proxima Centauri. The proportions are distorted for better understanding.*

Doppler effect

Level: first year of college

Consider an acoustic wave emitted by a fixed source S in the direction of a receptor moving away from the source at constant velocity v (Figure 41). Let c denote the wave velocity. The wave is supposed to be made of brief sound 'clicks' emitted with a periodicity T. We take as origin of time the emission of a first click. The receptor, located at point R, receives it at time t_1 given by:

$$t_1 = \frac{SR}{c}.$$

The next click is emitted at time T, during which the receptor has moved from R to R', so that it will be received at time t_2 such that:

$$t_2 = T + \frac{SR'}{c}.$$

The apparent period T', i.e., the time interval between two receptions by the receptor, is thus given by:

$$T' = t_2 - t_1 = T + \frac{SR' - SR}{c}.$$

Since $SR' - SR = vT'$, one obtains:

$$T' = \frac{T}{(1 - \frac{v}{c})}.$$

The apparent frequency is given by the inverse of the period, hence:

$$N' = N(1 - \tfrac{v}{c}).$$

For positive v (the receptor moves away from the source), the apparent frequency is lower than the true one.

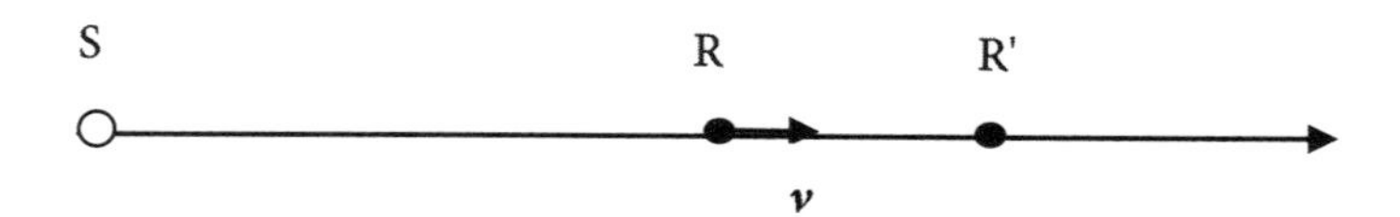

Fig. 41. *A fixed source emits two signals separated by a time interval T during which the receptor has moved from R to R'.*

Doppler-Fizeau effect

The wave velocity is now the velocity of light c. In principle, therefore, a relativistic calculation is required. However, when $v/c \ll 1$, the above result would still be a first approximation.

Coriolis force

Level: college

In a reference frame linked to Earth, the Coriolis force has the form:

$$\vec{F}_c = 2m\vec{v} \wedge \vec{\Omega}, \tag{A.2}$$

where $\vec{\Omega}$ is the rotation vector of Earth along the polar axis, with length $\frac{2\pi}{24 \times 3600}$, and $\vec{v}$ is the velocity vector of the moving mass m in the terrestrial frame of reference. One immediately sees that, due to the cross product, the force is zero in the case of a free fall at the poles, but is non zero for a pendulum having a trajectory perpendicular to the vertical axis.

Reich's experiment

This delicate experiment has been performed in 1831 in a deep mine shaft near Freiberg in Saxony. The theoretical deviation was 27.5 mm; Reich's measure was 28.3 mm. Let us calculate the eastward deviation for a falling mass m with no initial velocity.

Consider Newton's dynamic equation in the frame of reference linked to the Earth, including the weight and the Coriolis force:

$$m\frac{d^2\vec{r}}{dt^2} = m\vec{g} + 2m\vec{v} \wedge \vec{\Omega},$$

where $\vec{v} = \frac{d\vec{r}}{dt}$ denotes the velocity vector of the point mass m and the $\vec{g}$ acceleration of gravity along the local vertical axis. Upon integrating with respect to time, one gets:

$$\vec{v} = \vec{g}t + 2\vec{r} \wedge \vec{\Omega},$$

since the initial velocity is zero. Upon integrating once more, one obtains:

$$\vec{r} = \frac{1}{2}\vec{g}t^2 + \vec{\varepsilon}(t), \qquad (A.2)$$

with:

$$\vec{\varepsilon}(t) = 2\int_0^t \vec{r}(s)ds \wedge \vec{\Omega}.$$

In this last equation, one replaces $\vec{r}(s)$ by the dominant term in A.2, which corresponds to a first order perturbation calculation. This leads to:

$$\vec{\varepsilon}(t) = 2\vec{g} \wedge \vec{\Omega}\int_0^t \frac{1}{2}s^2 ds,$$

with modulus:

$$\varepsilon(t) = \frac{1}{3}gt^3\Omega\cos\theta,$$

where ϑ denotes the local latitude. Let t and h denote the duration and height of fall respectively, one has, to first order, $h = 1/2gt^2$, then:

$$\varepsilon(t) = \frac{2}{3}\Omega\cos\theta\sqrt{\frac{2h^3}{g}}.$$

Numerically, with the values h = 158.5 m, ϑ = 51° and g = 9.81 m/s², one recovers the value 27.5 mm.

Remarks:

1. The direction of the vertical is that of a plumb line which, being immobile, is not submitted to the Coriolis force.
2. In the experiment of a cannonball fired vertically, the total deviation would be zero: westward going up, and eastward going down.
3. The Coriolis force applies to atmospheric motions: ascending masses of air are deviated westward. Galileo's intuition that the origins of trade winds was in the rotation of Earth was correct, even though he did not have the theoretical framework to demonstrate it.

FOUCAULT'S PENDULUM

Consider a frame of reference attached to Earth, with the z-axis along the local vertical and the x-y plane tangent to the surface with the x-axis pointing eastward along a circle parallel to the equator and the y-axis pointing northward along a meridian circle. Then:

$$\vec{\Omega} = \begin{pmatrix} 0 \\ \Omega\cos\theta \\ \Omega\sin\theta \end{pmatrix} \text{ and } \vec{v} = \begin{pmatrix} \dot{x} \\ \dot{y} \\ 0 \end{pmatrix}$$

Taking into account the restoring force due to gravitation and the Coriolis force, Newton's dynamic equation, upon projection on the tangent plane, reads (in the limit of small oscillations):

$$\ddot{x} = -\omega^2 x + 2\dot{y}\Omega\sin\theta$$
$$\ddot{y} = -\omega^2 y - 2\dot{x}\Omega\sin\theta$$

with $\omega = \sqrt{\frac{g}{l}}$. Using a complex notation $z = x + iy$, the previous system of equations reads:

$$\ddot{z} + 2i\Omega\sin\theta\dot{z} + \omega^2 z = 0.$$

Introducing $z = e^{rt}$, one obtains:

$$r^2 + 2i\Omega r\sin\theta + \omega^2 = 0.$$

with the solutions:

$$r = -i\Omega\sin\theta \pm i\omega_0.$$

where $\omega_0^2 = \Omega^2 \sin^2\vartheta + \omega^2$. Since $\Omega \ll \omega$ (compare $\frac{2\pi}{16}$ to $\frac{2\pi}{24\times 3600}$), one can replace ω_0 by ω. Then:

$$z(t) = e^{-i\Omega t\sin\theta}[ae^{i\omega t} + be^{-i\omega t}],$$

with:

$$z(0) = a + b$$

and:

$$\dot{z}(0) = -(a+b)i\Omega\sin\theta + i\omega(a-b).$$

Assume that the pendulum is set in motion along the x-axis without initial velocity $(\dot{z}(0)=0)$, which implies that $a + b$ is a real number and $(a+b)\Omega\sin\theta = (a-b)\omega$. Then:

$$a = b\frac{\omega+\Omega\sin\theta}{\omega-\Omega\sin\theta} \simeq b\left(1+2\frac{\Omega}{\omega}\sin\theta\right) \simeq b.$$

and finally:

$$z(t) \simeq z(0)e^{-i\Omega t\sin\theta}\cos\omega t.$$

The motion is approximately taking place in a plane rotating with period:

$$T = \frac{2\pi}{\Omega\sin\theta}.$$

APPENDIX B

Kelvin's Model and Calculation

METHOD

It is assumed that at t = 0 Earth is a homogeneous sphere the temperature of which is uniform and close to that of melting rocks (around 3900°C); it cools down by conduction of heat towards the surface, and convection is not considered. The time evolution of the temperature T is then given by Fourier's heat equation:

$$\frac{\partial T}{\partial t} = \kappa \Delta T, \tag{B.1}$$

where κ denotes the ratio of thermal conductivity to specific heat of the rock, t the time, and Δ the Laplacian operator. The initial conditions are known; the boundary conditions are imposed (the temperature at Earth's surface), and the function $T(t,\vec{r})$ can then be calculated, as well as its gradient at Earth's surface $r = R$, which evolves with time. Since this gradient is empirically known to be equal to 30°C per kilometer, the age of Earth is given by the time it took for the gradient to decrease from its infinite initial value to the actual value.

Calculation

The actual value of the surface thermal gradient is such that the limiting temperature of 3900°C is reached within less than 200 kilometers inward. The region where the temperature varies appreciably is thus small compared to the radius of the planet, so that one can reduce the problem to a one-dimensional one, perpendicular to Earth's surface. Under this approximation, one can show that the surface thermal gradient is given by:

$$\left.\frac{dT}{dx}\right|_{x=0} = \frac{T_0}{\sqrt{t\kappa\pi}}$$

Demonstration
Let us introduce the Fourier transform with respect to x as:

$$\Im(T) = \tilde{T}(k,t) = \int T(x,t)e^{-ikx}dx.$$

The Fourier transform of the second derivative of T is equal to $F(T'') = -k^2F(T)$. Taking the Fourier transform with respect to x of Eq. (B.1) leads to:

$$\frac{\partial \tilde{T}(k,t)}{\partial t} = -k^2\kappa\tilde{T}(k,t),$$

with the solution:

$$\tilde{T}(k,t) = \tilde{T}(k,0)e^{-k^2\kappa t}.$$

We now have to take the inverse Fourier transform of this solution. Since the Fourier transform of a convolution product of two functions is equal to the ordinary product of the Fourier transforms of the functions, one can write directly:

$$T(x,t) = T(x,0) * \Im e^{-k^2\kappa t}.$$

The inverse Fourier transform of a gaussian function is still a gaussian function. Taking into account the normalization factor, one gets:

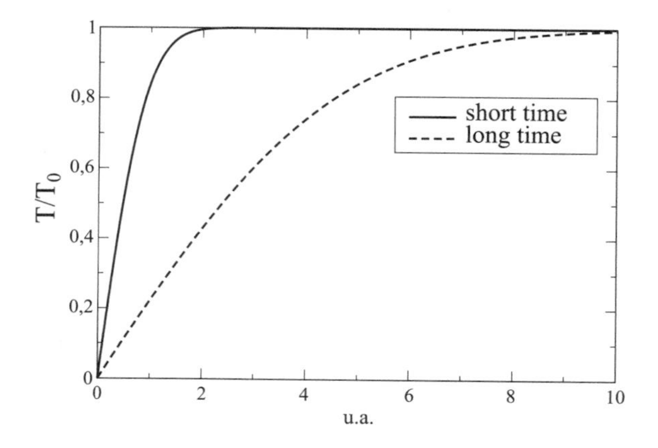

Fig. 42. *Variation of temperature with distance from the surface of Earth (in arbitrary units), for two different times. The thermal surface gradient decreases with time.*

$$T(x,t) = \frac{1}{2}\sqrt{\frac{1}{\pi\kappa t}} e^{-\frac{x^2}{4t\kappa}} * T(x,0).$$

Kelvin chose as initial condition $T(x,0) = T_0$ for all x, and for limiting condition[2] $T(0,t) = 0$ for all t. It is then easy to show that the choice $T(x,0) = T_0[Y(x) - Y(-x)]$ satisfies both conditions. $Y(x)$ denotes the Heaviside function, equal to 0 for $x<0$ and equal to 1 for $x>0$. An elementary calculation then leads to:

$$T(x,t) = \frac{2T_0}{\sqrt{\pi}} \int_0^{\frac{x}{2\sqrt{t\kappa}}} e^{-z^2} dz.$$

Equation B.2 for the thermal surface gradient is then obtained upon derivation with respect to x and setting $x=0$. In his original paper (Thomson, 1863b), Kelvin takes a value of 0.36°C per meter, an initial temperature of 3900°C and a value for κ of $1.2 \times 10^{-6}\ m^2 s^{-1}$, which, given the uncertainties on the data, leads to a range:

$$20 < t < 400 \text{ million years.}$$

APPENDIX C

Radioactivity

LEVEL : FIRST YEAR OF COLLEGE

Lord Kelvin did the best he could with the knowledge of his time, but he could understand neither the origin of the solar energy nor the origin of the internal temperature of Earth. Both phenomena are now explained within theories which were unknown at the time: relativity theory and nuclear physics. The key argument is the following: nuclear transformations take place both in the Sun and in Earth, which imply a mass defect Δm accompanied by an equivalent thermal energy production, according to Einstein's famous equation:

$$E = \Delta mc^2$$

where c denotes the velocity of light. As an example, one gram of mass defect per second corresponds to an energy power of 9×10^{13}W.

This mass defect has two different nuclear origins:

- In the case of Earth, heavy elements transform into lighter ones. Fission is one of the mechanisms taking place, also used in nuclear plants and atomic bombs. A fissile nucleus breaks into two nuclei of similar masses. The other process is a disintegration by which a heavy nucleus emits an α-particle (a helium nucleus), or an electron. It is these processes which are used for dating.
- In the case of the Sun, hydrogen nuclei (protons) undergo a series of fusion reactions leading to a helium nucleus. Fusion reactions are also used in hydrogen bombs, and fusion reactors for producing electricity are under investigation (starting with deuterium and tritium, two heavy isotopes of hydrogen).

What is the origin of solar energy?

It is often said that, in the Sun, hydrogen "burns" into helium. This should not be understood as a classical combustion, which is a chemical reaction involving interactions between electrical charges – electrons and nuclei. Here, it is a process by which the nuclei transform into other nuclei through the effects of the strong and weak nuclear forces. For the same amount of matter, a nuclear transformation involves an energy 10 to 100 million times larger than a chemical reaction.

The sequence of fusion reactions leading to the synthesis of a helium nucleus is the following:

$$p + p \rightarrow d + e^{+} + \nu$$
$$d + p \rightarrow {}^{3}\mathrm{He}$$
$${}^{3}\mathrm{He} + {}^{3}\mathrm{He} \rightarrow {}^{4}\mathrm{He} + 2p$$

The net effect is the conversion of 4 protons into a helium nucleus:

$$4p \rightarrow {}^{4}\mathrm{He} + 2\nu + 2e^{+} + 26.7\ \mathrm{MeV}$$

The positrons (e^{+}) annihilate with electrons into photons. The neutrinos (ν), interacting very weakly with matter, escape from the star with little energy. Most of the energy appears as kinetic energy of the fusion products, which thermalizes through collisions: hence the name *thermo*nuclear reactions.

In the center of the Sun, each second, 624.5 million tons of hydrogen transform into 620 million tons of helium. This loss of mass of 4.5 million tons per second corresponds, via Einstein's relation, to a power of 4×10^{26} Watts. It is believed that, in the course of its life, about one thousandth of the solar mass will have transformed into helium. Taking into account the mass of the Sun (2×10^{30} kg), its overall lifetime will therefore be 10 billion years, i.e., twice its present age.

If the origin of the solar energy was of chemical nature, for example, the recombination of electrons and protons to form hydrogen atoms, each atom would release 13.6 eV, compared to 26.7 MeV in the nuclear case. Taking into account that 4 protons are required for a helium nucleus, the ratio of energies involved in both cases is $1/4\times26.7\times10^{6}/13.6 \approx 50{,}000$. The expected lifetime of the sun would then be 20,000 years, way too short: Kelvin knew about that!

What is the origin of the internal heat of the Earth?

Most of the energy (80 per cent) is of nuclear origin and due to the disintegration of elements with large half-life times. The initial heat, due to the gravitational collapse or to the radioactivity of elements with short half-life times (as ^{26}Al) has been dissipated a long time ago. A small fraction of the energy is due to the solidification of the central part of the planet, which liberates the corresponding latent heat.

Radioactive transformations concern essentially elements with long half-life times (larger than one billion years): uranium-235 and uranium-238, disintegrating into lead-207 and lead-206 respectively; thorium-232 disintegrating into lead-208; and potassium-40, disintegrating into argon-40 and calcium-40.

The geothermal flow of energy of Earth is $4.6x10^{13}$ W. Given the total area of the planet (5×10^{14} m^2), this geothermal flux is 0.1 W/m^2 only, compared to the average solar flux of 342 W/m^2. This implies that the surface temperature of Earth results from the solar radiation. In the presence of the geothermal flux only, the temperature of Earth would be –243°C.

Radiochronology

Level: first year of college

The various forms of de-excitation of an atomic nucleus, whether it be fission or radioactivity, are phenomena pertaining to the quantum behavior of matter. These are basically random processes, for which predictions can still be made using statistical laws of large numbers. It is therefore possible to describe the behavior of an ensemble of radioactive nuclei. The average number of nuclei transforming in a unit of time is simply proportional to the number of existing nuclei $N(t)$ at time t:

$$\frac{dN(t)}{dt} = -\lambda N(t),$$

where λ depends on the element considered.

This law can be deduced from the following physical hypotheses: i) all nuclei are identical, ii) the probability of transformation of a nucleus is independent of its age, and iii) the transformation of a nucleus has no influence on the others. From the above equation, one can deduce the law of evolution of the population $N(t)$:

$$N = N_0 e^{-\lambda t}, \tag{C.1}$$

where N_0 denotes the initial average number of nuclei.

The period, or half-life, T of the element is defined as the time interval necessary for the number of nuclei to be divided by 2:

$$N(t+T) = \frac{1}{2} N(t)$$

From $e^{-\lambda T} = 1/2$, one gets $T = \mathrm{Ln}(2)/\lambda$, a value independent of t. This property is characteristic of the exponential function.

It is remarkable that the half-lifes of the various elements cover a huge range of values, extending to 25 orders of magnitude. For example, the half-life of Beryllium-8 is 10^{-16} years; it is 5,730 years for carbon-14, and 50 billion years for rubidium-87.

In the case of the longest half-lifes, one can wonder whether the number of disintegrations is large enough to be observable. Let us consider the example of rubidium-87. The number of nuclei in 87 grams of that element is equal to Avogadro's number, $N_0 = 6.02 \times 10^{23}$. The value of λ corresponding to a half-life T of 50 billion years is given by $\mathrm{Ln}(2)/T$, i.e., $4.4 \times 10^{-19}\ \mathrm{s}^{-1}$. The number of disintegrations during one second of time is therefore:

$$\Delta N \simeq \lambda N_0 = 2.7 \times 10^5,$$

which amounts to 3 per second if one starts with a mass of 1 mg only: this is easily detectable.

Let us now describe the method of dating. Several cases have to be considered:

1. The initial population N_0 is known. After an unknown time t, one measures the number of nuclei $N(t)$. The time t is determined by use of the formula:

$$t = \frac{1}{\lambda} \log \frac{N_0}{N(t)},$$

deduced from eq. C.1.

How can N_0 be determined?

This is the case of carbon-14 dating. The isotopic ratio carbon-14 to carbon-12 of the atmosphere is nearly constant over time:[3] carbon-14 is unstable, but it is renewed by the action of cosmic rays on nitrogen-14 in the upper atmosphere. Through photosynthesis, the same ratio is found in all the plants, and also in all living organisms eating plants, thus renewing their carbon. A human adult experiences about 10,000 disintegrations per second, half of them due to carbon-14 and the other half to potassium-40. When a living organism dies, its carbon-14 is no longer renewed, it simply disintegrates. It should be borne in mind that this method can only be used for historical times, since carbon-14 has a half-life of 5,730 years and after a few half-lives, not enough nuclei remain to be measured with good accuracy. For dating geological events, we have seen that the lead-lead method requires the knowledge of the initial populations.

2. N_0 is not known, but the number of descendant nuclei N_d is zero at time $t = 0$. One can then write:

$$N_d(t) = N_0 - N(t) = N(t)e^{\lambda\tau} - N(t) = N(t)(e^{\lambda t} - 1) \qquad \text{C.2}$$

or

$$t = \frac{1}{\lambda} Ln\left(\frac{N_d(t)}{N(t)} - 1\right)$$

3. Similar situation, however one does not assume $N_d(0) = 0$. Eq. C.2 now writes:

$$N_d(t) = N(t)(e^{\lambda t} - 1) + N_d(0)$$

Consider the case of the disintegration ^{87}Rb ^{87}Sr. Denoting concentrations with brackets, one writes:

$$[^{87}\mathrm{Sr}] = [^{87}\mathrm{Rb}](e^{\lambda t} - 1) + [^{87}\mathrm{Sr}]_0.$$

Strontium in the mineral is present under two isotopic forms: strontium-86 and strontium-87. Strontium-86 is not of radiogenic origin, therefore $[^{86}\mathrm{Sr}]=[^{86}\mathrm{Sr}]_0$. Strontium-87 has two origins: it is present at the initial time, and it is also continuously produced from rubidium-87.

Experimentally, it is easier to determine isotopic ratios than absolute concentrations, so one introduces the concentration in strontium-86, which is constant. One then has:

$$\frac{[^{87}\mathrm{Sr}]}{[^{86}\mathrm{Sr}]} = \frac{[^{87}\mathrm{Rb}]}{[^{86}\mathrm{Sr}]}(e^{\lambda t} - 1) + \frac{[^{87}\mathrm{Sr}]_0}{[^{86}\mathrm{Sr}]_0}.$$

In this equation, both the time t and the initial concentration in strontium-87 are unknown. How can one determine two quantities from one equation? The hint consists in noting that the proportion of rubidium to strontium varies in different samples of the same mineral. The initial isotopic ratio of the two isotopes of strontium will however be the same, since their chemical properties are the same. In other words, if the above equation is written in the form:

$$y = ax + b$$

with

$$y = \frac{[^{87}Sr]}{[^{86}Sr]} \quad \text{measured for each sample}$$

$$x = \frac{[^{87}Rb]}{[^{86}Sr]} \quad \text{measured for each sample}$$

and

$$a = e^{\lambda t} - 1 \approx \lambda t \quad \text{unknown}$$

$$b = \frac{[^{87}Sr]_0}{[^{86}Sr]_0} \quad \text{unknown,}$$

upon analyzing different samples of the same mineral, one has access to different values of the couple (x,y).

One checks that the experimental points lie on a straight line, which validates the method. The slope of the line is directly related to the age of the mineral since closure. Knowing λ, one determines t (and b). Figure 43 shows the result of such an analysis for meteorite Allende.[4]

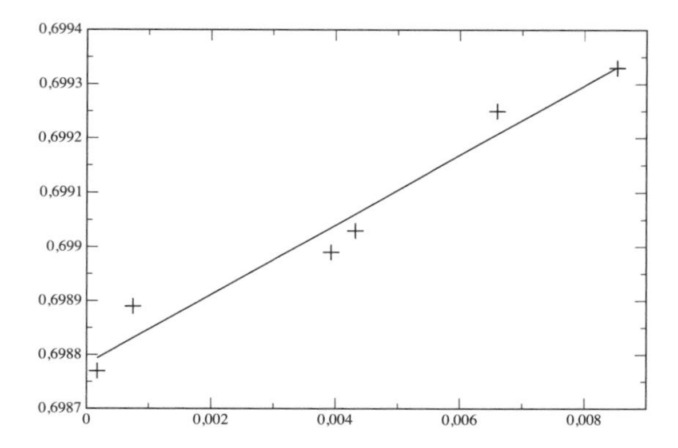

Fig. 43. *Isochrone method applied to the meteorite Allende. The equation of the best straight line is: $y = 0.06426x + 0.6988$*

The points are taken from (Gray, 1973). The half-life of rubidium-87 is 48.8 billion years, leading to a value $\lambda = 1.42 \times 10^{-11}$ $year^{-1}$. The slope is found to be a = 0.06426, hence the age of the meteorite is:

$$t = \frac{\text{Ln}1.06426}{1.42 \times 10^{-11}} = 4.39 \times 10^{9} \text{ years.}$$

The method is called 'isochrone', since the straight line determined by the experimental points corresponds to a given elapsed time.

Let us recall that these different methods all assume that the sample has been a closed system since the initial time, taken here as time 0, in the sense that no addition or loss of matter used for dating has taken place. This condition is fulfilled for a mineral since the moment of its solidification (no internal migration); in the case of carbon-14 dating, since the death of the living organism (no metabolism).

APPENDIX D

The Copernicus/Tycho Brahe Equivalence

LEVEL: FIRST YEAR OF COLLEGE

We shall derive a formal equivalence of two models in their simplified versions: the heliocentric model of Copernicus and the geocentric model of Tycho Brahe.

We shall make three assumptions:

- the planet orbits are perfect circles,
- the data used by both astronomers satisfy Kepler's third law, i.e., the square of the period E (around the Sun) is proportional to the cube of the distance b to the Sun,
- the orbits are coplanar.

These three assumptions are good approximations and make the calculations much simpler.

In the system of Copernicus, the planets circle around a fixed sun. The equations of the trajectories can be written as:

$$x = b\cos\frac{t}{b^{3/2}} \qquad \text{(D.1)}$$

$$y = b\sin\frac{t}{b^{3/2}}$$

For simplicity, we have taken, for Earth, $b = 1$ and $E = 2\pi$. The factor $b^{3/2}$ comes from Kepler's third law.

In the system of Tycho Brahe, all the planets (except Earth taken as origin) circle around the Sun.

All the other planets have the same deferent circle with radius unity (distance Earth–Sun ES) and each epicycle has a radius b equal to the distance between the planet and the sun. The equations of motion are derived from D.1 upon translating the motion of Earth:

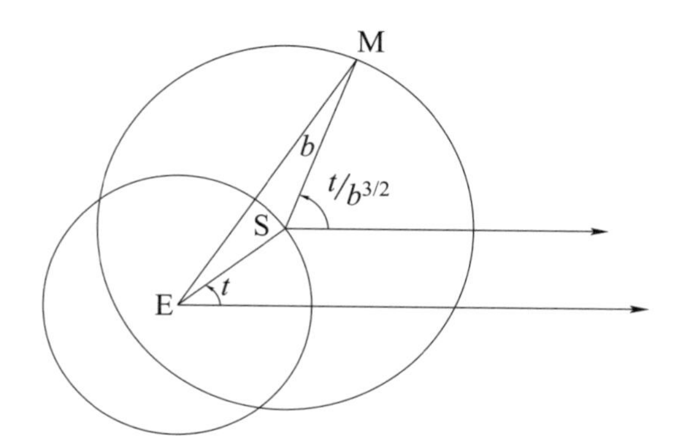

Fig. 44. *Schematic motion of Mars according to Tycho Brahe*

$$x = -\cos t + b\cos\frac{t}{b^{3/2}} \qquad \text{(D.2)}$$
$$y = -\sin t + b\sin\frac{t}{b^{3/2}}.$$

It is the same trajectory as in the system of Copernicus, but seen from Earth. Eq. D.2 does represent correctly the retrograde motions,[5] as can be seen from Figure 45. The comparison with Kepler's sketch is striking.

Note: Formula D.2 can be rewritten as:

$$x = b\cos\frac{t}{b^{3/2}} - \cos t$$
$$y = b\sin\frac{t}{b^{3/2}} - \sin t$$

which can be interpreted à la *Ptolemaus* (one recovers a fixed Earth taking $b = 1$). For the external planets ($b > 1$), the first term is dominant.

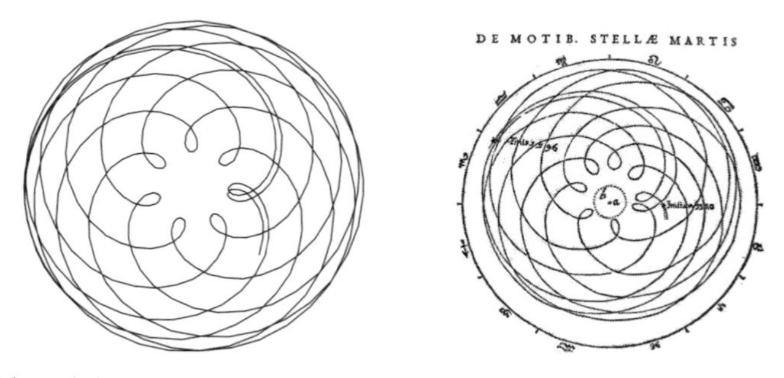

Fig. 45. *Plot of the trajectory of Mars as given by Eq. D.2 (left), using a computer, compared to the sketch of Kepler (right), made five centuries ago.*

When $b \gg 1$, the orbit is close to a circle (Neptune has $b = 30$), and the second term appears as a small modulation of the first. The radius of the epicycle is smaller than the radius of the deferent. Retrograde motions being phenomena with little angular amplitude, it was quite natural to take them into account by a small correction to the principal motion.

For the five planets known by Tycho Brahe, Mercury, Venus, Mars, Jupiter and Saturn, the values of b are 0.39, 0.72, 1.52, 5.2 and 9.55 respectively.

APPENDIX E

The Relativity of Trajectories

LEVEL: FIRST YEAR OF COLLEGE

We shall address here the observation made on page 141, according to which a trajectory has a meaning only with respect to a given frame of reference. Let us examine how one can construct a cycloïd (trajectory with respect to the road of the valve of a bicycle tire) and the trochoïd (trajectory of a nail of length h).

The road is taken along Ox; the hub of the wheel of radius r is Ω. Point M represents the valve on the wheel for the cycloïd, and the end of the nail for the trochoïd. Let t be the angle of rotation of the wheel. Rolling without slipping implies that $OH = rt$.

Equation of the cycloïd

The simplest is to consider the complex plane. Let z be affix of M.

Then:

$$z = rt + ir + re^{i(\frac{3}{2}\pi + t)}.$$

which can be written as:

$$z = rt - r\sin t + ir(1 - \cos t).$$

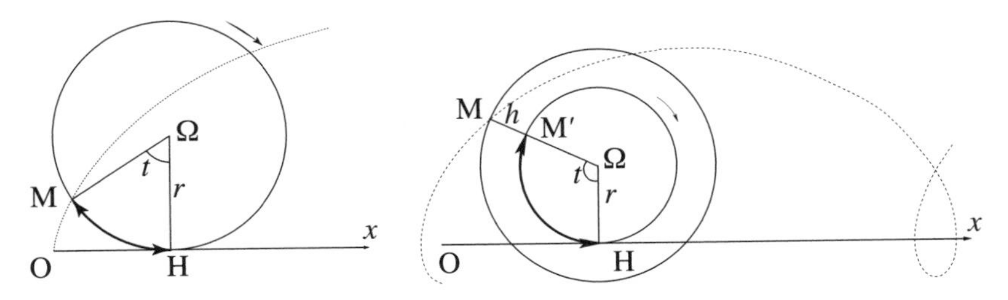

Fig. 46. *Construction of the cycloïd (left), and of the trochoïd (right)*

A parametric equation of the cycloïd follows:

$$x = r(t - \sin t)$$
$$y = r(1 - \cos t).$$

Equation of the trochoïd

The only difference with the previous calculation is that point M is at a distance $r + h$ of the center Ω. Hence:

$$z = rt + ir + (r + h)e^{i(\frac{3}{2}\pi + t)}.$$

The equation of the trochoïd follows:

$$x = r(t - \sin t) - h \sin t$$
$$y = r(1 - \cos t) - h \cos t.$$

Acknowledgements

'The philosophy of science is as useful to scientists as ornithology is in teaching birds to sing.' This aphorism of Feynman, one of the greatest physicists of the twentieth century, is accepted with amused unanimity by professionals in the hard sciences, for whom philosophers and sociologists of science – inasmuch as they are familiar with them – enjoy little prestige. If it is trivially correct, this sally deserves a riposte: ornithology is not useful to birds; it is useful to people. This is also what one is entitled to expect from the philosophy of science. For the reasons indicated in the Preface, this expectation is often disappointed – at least in France. And I owe it to Jacques Bouveresse to have brought his talent as a philosopher (he really is not very orthodox) to the service of a cause that may seem banal to those odd birds that physicists are: it is always objective reality, that which science postulates and seeks to discern, that in the last analysis makes it possible to test the various hypotheses and representations of the world. If everything were a matter of convention, of balance of forces and networks, then we would see less funding given to research and more to communication …

Despite being signed by a single author, this book has drawn in many places not only on the ideas of Xavier Campi and Jacques Treiner, but also on their writings.

I have benefited both from Jacques's initial contribution, and throughout from his criticisms, commentaries, questions and disagreements, which have always been very productive.

Large parts of the book are the result of a close collaboration with Xavier; his knowledge of astronomy both ancient and modern has been most important.

Several other colleagues and friends have contributed their knowledge. This socialization of the book's development was made indispensable by the scope of knowledge needed (in astronomy, geophysics, biology, religions, philosophy, history, geography...) in order to cope with so broad a subject.

Thanks accordingly to: Marc Alizon, Daniel Becquemont, Jean-Pierre Bibring, Claude Birman, Oriol Bohigas, Hubert Bost, Jean Bricmont, René Cori, Philippe Depondt, Philippe Deterre, Ahmed Djebbar, Denise Douzant, Jean Duprat, Egidio Festa, Pierre Fiala, Alexandre Ghazi, Roger Hewins, Thierry Hoque, Catherine Jami, Vincent Jullien, Jean-Louis Krivine, Nabil Lemrazi, Bruno Levallois, Tony Lévy, Ilana Löwy, Satya Majumdar, Charles Malamoud, Farouk Mardam-Bey, Sylvie Nony, Nicolas Pavloff, Bernard Raoult, Pascal Richet, Nicolas Sator, Otto Schaefer, Youssef Seddik and Brigitte Zanda.

This collective thanks does not take into account the hours of exciting discussion I have had the pleasure of sharing with many of these contributors, for example on Islamic-Arab and Jewish culture, the subtleties of Christian exegesis, the contributions of Galileo or the physico-chemical conditions for the origin of Earth.

Irène, Léna, Jean-Michel, Catherine, Béatrice, Bernard, Sonia and Alain kindly helped me to correct all or part of the text.

Françoise Gicquel and Thérèse-Marie Mahé courageously took on a different kind of translation: from physics into French. This book owes much to them.

Marcel Vénéroni's help and his presence have been unclassifiable and valuable. My thanks to him.

I express my gratitude to the staff of the libraries of Paris-Sud, Paris VI, the Sorbonne, Sainte-Geneviève, the Institut du Monde Arabe, the Institut Catholique de Paris (particularly to Gwladys Gouttegas), the library of Le Saulchoir and the Bibliothèque Nationale de France for their skill and availability.

I appreciated the welcome I received from the Laboratoire de Physique Théorique et Modèles Statistiques (LPTMS) at the university of Paris-Sud (Orsay), the Laboratoire de Physique Nucléaire et de Hautes Énergies (LPNHE) of the Université Pierre et Marie Curie and the Espace Pierre-Gilles de Gennes of the ESPCI Paris Tech.

Finally, there was the admirable work of André Bellaïche and Éditions Cassini, who were able to metamorphose a manuscript into a book.

If the danger of original ideas is error,[1] the reader will be readily convinced that from this point of view I have taken few risks. Keen to maintain respectability and the friendship of the colleagues who helped me in this work, I accept the paternity of the more doubtful 'originalities' that may none the less still persist. They are neither responsible nor guilty for these.

Notes

Foreword

1 Nikkie R. Keddie, *An Islamic Response to Imperialism* (University of California Press, 1968), quoted by Professor Irfan Habib in his paper 'Islam, Colonialism and Modern Science during the Nineteenth Century', presented at the Global Economic History Network (GEHN) Conference-Taiwan, 9–11 May 2006.

Introduction

1 This modern rejection of science is often accompanied by a revival of religion. We shall not discuss here the wide subject of the causes of religion, on which atheist writers are generally in agreement. Here is a brief sample of these.

Laplace (1995) had a rather optimistic view when he wrote in 1814:

> but these imaginary causes have successively retreated along with the limits of our knowledge, and they disappear entirely before healthy philosophy, which sees in them only the expression of our ignorance of real causes … all these extraordinary phenomena were seen as so many signs of heavenly wrath. People appealed to heaven to turn away their harmful influence. They did not pray to suspend the course of the planets or the Sun: observation would soon have shown the uselessness of such prayers.

Laplace's dialogue with Napoleon is also famous:

> 'Monsieur de Laplace, I cannot find in your system any mention of God.' 'Sire, I had no need for that hypothesis. … That hypothesis, Sire, in fact explains everything but makes it possible to predict nothing. As a scientist, I am obliged to offer you work that makes possible predictions.'

Though Marx's famous definition of religion as the 'opium of the people' (Marx, 1975) is well known, the full quotation is rather more subtle:

> *Religious* distress is at the same time the *expression* of real distress and also the *protest* against real distress. Religion is the sigh of the oppressed creature, the heart of a heartless world, just as it is the spirit of spiritless conditions. It is the *opium* of the people.

Nor was Freud far from this in 1927 when he wrote in *The Future of an Illusion* (Freud, 1961):

> But man's helplessness remains and along with it his longing for his father, and the gods. The gods retain their threefold task: they must exorcise the terrors of nature, they must reconcile men to the cruelty of Fate, particularly as it is shown in death, and they must compensate them for the sufferings and privations which a civilized life in common has imposed on them.

It is the unexpected rise in the grip of religion and the irrational in the twenty-first century that concerns us, not its endemic existence. For a detailed study of this phenomenon, Bouveresse (2007) will offer interesting and even pleasurable reading. Among the rare studies of atheism, see the very well-documented work of Georges Minois (1998).

2 Which is already strange. It is the characteristic of scientific ideas to be verifiable, and thus debatable. In science no ideas are irrefutable; there are only ideas that have not been refuted.
3 Jacques Bouveresse, *Prodiges et vertiges de l'analogie. De l'abus des belles-lettres dans la pensée* (Raisons d'Agir, 1999).
4 In a more tragic context, we may recall, science became 'Aryan' under Hitler and 'proletarian' under Stalin.
5 For example, in India some people champion a return to 'Vedic' science in opposition to supposedly 'Western' science.
6 The USA is a rich country, but we see the survival or even growth of religious sects there.
7 This is not new. The 'sabotage' of the nineteenth century, literally throwing a clog (*sabot*) into the works to break the machine, has been viewed as a way of opposing 'scientific' progress.
8 A regression even in relation to the golden age of Islamic civilization.
9 Even when they choose to specialize in science, many students do so because of the more promising employment possibilities rather than out of intrinsic interest.

How to Use This Book

1 One need only note the disproportionate presence on the internet of creationist and fundamentalist sites, often concealed by masks of a perfectly respectable scientific window-dressing.

Part One: Earth's Age

1 A summary of this history, adapted for secondary education, can be consulted in Hubert Krivine et al., Âge *de la terre.*
2 Curiously called since the 1960s 'falsifiable', after the English edition of Karl Popper's *Logic of Scientific Discovery*. For the French edition, Popper preferred '*réfutable*'. Even in English, 'falsifiable' was a neologism.
3 Steven Weinberg, *The First Three Minutes*.
4 Stanley Awramik, 'Respect for stromatolites', and Fabienne Lemarchand, 'Controverse: les premières traces de vie'.
5 This is particularly true for the age of the universe, a subject on which empirical data and theories are still far from unanimous.
6 And that of human beings? As we shall see at the end of this Part, relative uncertainty as to the date of Earth's origin is of almost the same order of magnitude.
7 The *Pour la science* dossier gives a good summary of our knowledge of dating in general, from a billion years ago to the human scale.
8 Étienne Klein, *Discours sur l'origine de l'univers*.

1. 'Pre-Science'

1 Jean-Pierre Vernant, *L'univers, les dieux, les hommes* (Seuil, 1999).
2 Paul Veyne, *Writing History: An Essay on Epistemology* (Manchester University Press, 1984).
3 Every philosophy student who has to comment on this text raises the question: but why this sudden *clinamen*? It is no longer the right question: today we know that many phenomena appear as the result of small random fluctuations – what are called spontaneous symmetry breaking. What is naïve on the part of Lucretius is the assumption of pre-existing weight 'at the beginning'. What is 'up' and what is 'down'? If there was nothing, what is the meaning of 'falling' (in relation to what?): the uniform velocity could just as well have been zero …

4 The thesis defended in Karl Marx, 'Difference between the Democritean and Epicurean Philosophy of Nature', in *Marx/Engels Collected Works*, vol. 1 (Lawrence & Wishart, 1975), in 1841 and in his notebooks on Epicurus contains a reflection on this contention, and particularly a mine of Greek and Latin references. What is interesting here is not his very embryonic Marxism, but his erudition.

5 Aristotle's disciples were the 'peripatetics' who studied in the Lyceum while the master walked about (Greek: *peripatein*).

6 Gouguenheim (Sylvian Gouguenheim, *Aristote au Mont Saint-Michel. Les racines grecques de l'Europe chrétienne* [Seuil, 2008]) has recently challenged this connection, though his opinion has only marginal support among historians (Max Lejbowicz, 'Aristote au Mont Saint-Michel. Les racines grecques de l'Europe chrétienne'. *Journal of Medieval Studies*, 2008).

7 And still later; this is what His Holiness Leo XIII wrote as late as 1889 (Albert Farges, *Études philosophiques pour vulgariser les theories d'Aristote et de saint Thomas et montrer leur accord avec les sciences* [Bertche et Tralin, 1909]): 'The further you proceed on this path, the more you will be secured and strengthened in the conviction that Aristotelian philosophy, as interpreted by St Thomas, rests on the most solid foundations, and that this is where the most certain principles of the most solid and useful science are still to be found today ...'

8 A side effect of this conception was the unwillingness of academic philosophers to recognize the existence of meteorites, which violated the immutability of the supralunary world. It was only with the work of Jean-Baptiste Biot in 1803 (Nicolas Witkowski, 'J.-B. Biot: un homme, une météorite', *La Recherche* 326, December 1999) that 'firestones' were no longer viewed as illusions peddled by uncouth and drunken peasants. This is an argument often used by champions of flying saucers: 'official' science refuses to see anything that upsets it. We shall return to this point in Part Three of this book.

9 Frank Michaeli, *Textes de la Bible et de l'ancien Orient* (Delachaux & Niestlé, 1961); Israel Finkelstein and Neil Asher Silberman, *The Bible Unearthed: Archaeology's New Vision of Ancient Israel and the Origin of Its Sacred Texts* (Free Press, 2001).

10 Cited by Pascal Richet, in Pascal Richet, *A Natural History of Time* (University of Chicago Press, 2009).

11 Averroes is referring here to sura 11, verse 7.

12 Averroes, *The Philosophy and Theology of Averroes. Tractacta* (Manibhai Mathurbhal Gupta, 1921).

13 Sylvie Nony, private communication, 2010a.

14 Compare this with the two initial purposes of the internet in the 1970s, which were also quite modest: a) to make possible rapid exchange of information between scientists at the Stanford Institute and UCLA; b) to ensure communication in the US military that would resist thermonuclear attack. Nothing more!

15 Considered the equal of his contemporaries Galileo or Descartes as a founder of modern scientific thinking, and not to be confused with Roger Bacon, the English philosopher of the thirteenth century, Francis Bacon never accepted the Copernican model, which he saw as no more than a mathematical game.

16 *Novum Organum*, book I, cxxix.

17 Vincent Julien, *Sciences 'Agent double'* (Stock, 2002); Gérard Simon, *Sciences et histoire* (Gallimard, 2008).

18 Here is a good summary of Newton's method as described by Jacques-Joseph Champollion (the older brother of the famous Jean-François) in his *Résumé complet de chronologie* published in 1830:

> Newton, who combined great piety with great knowledge, undertook in his free time to bring chronology into conformity with the order of nature, as he put it, with astronomy and sacred history, and to free it from all contradictions. He based his deductions on two principles: 1) the ancients estimated that three generations of men came to a hundred years; allowing thirty-three years for each generation, he accordingly reduced these for the generations or successions of kings to eighteen years each; 2) comparing the position occupied by the cardinal points in the sphere attributed to Chiron at the time of the Argonauts with the position in which Meto observed them in 432 BCE, and applying the principle of the precession of the equinoxes with a difference of seven degrees traversed against the order of the (zodiacal) signs from Chiro to Meto, he fixed the date of the Argonauts' expedition to the year 936 BCE; the other eras in Greek and Oriental history followed from this initial determination … and the capture of Troy was in 904 BCE.

19 As the Rev. James Ussher (1586–1656) put it: 'In the beginning God created Heaven and Earth (Gen. 1, v.). Which beginning of time, according to our Chronologie, fell upon the entrance of the night preceding the twenty third day of October, in the year of the Julian Calendar, 710 …' (James Ussher, *The Annals of the world, deduced from the origin of time and continued to the beginning of the emperour Vespasians reign, and the totall destruction and abolition of the Temple and commonwealth of the Jews, containing the historie of the Old and New Testament, with that of the Macchabees, also all the most memorable*

affairs of Asia and Egypt and the rise of the empire of the Roman Caesars under C. Julius and Octavianus ... [J. Crook and G. Bedel, 1658]). The Julian calendar, for Ussher, began on 1 January 4713 BCE. The Reverend Ussher is probably the most frequently cited biblical chronologist, since his date was still mentioned in the Authorized Version of the English Bible until the early twentieth century. A copious literature discusses Ussher (or Usher), a particularly good source being Gorst, *Measuring Eternity: The Search for the Beginning of Time* (Broadway Books, 2001). Also attributed to him, but wrongly, is the timing of the origin of the world to nine o'clock in the morning. This detail (Patrick Wyse Jackson, *The Chronologers' Quest: The Search for the Age of the Earth* [Cambridge University Press, 2006]) was in fact the achievement of his colleague John Lightfoot (1602–75).

20 A qualification should be made. Newton wrote in *The Chronology of Ancient Kingdoms Amended* (1725): 'I have drawn up the following Chronological Table, so as to make Chronology suit with the Course of Nature, with Astronomy, with Sacred History, with *Herodotus* the Father of History, and with it self; without the many repugnancies complained of by *Plutarch*. I do not pretend to be exact to a year: there may be Errors of five or ten years, and sometimes twenty, and not much above.'

21 The only ambiguity arose from the choice between two versions of the Bible, the Septuagint and the Vulgate. The Septuagint, which prevailed for more than a thousand years, claimed to be the Greek translation of the Hebrew Torah. According to legend, in the third century BCE, seventy sages, translating separately, reached an identical text: hence its name. The Vulgate, a new translation by St Jerome (347–419), was adopted by the Council of Trent in 1546; it remains more or less the official text of the Catholic church today. We should note that the Hebrew Bible, though in principle the basis of both variants, dates the Creation at 3760 BCE.

22 John G. C. M. Fuller, 'The Age of the Earth: From 4004 BC to AD 2002'. *Geological Society*, London, Special Publications 190, 2001.

23 Martin Gorst, *Measuring Eternity: The Search for the Beginning of Time* (Broadway Books, 2001).

24 Isaac De La Peyrère, *I preadamiti* (Quodlibet, Macerata, 2004).

25 For La Peyrère, who cites St Paul: 'God will hence wipe out the sins of the Jews. That means, he will remove before all else the lack of faith in the heart of the Jews. He will change their heart of stone into a heart of flesh.' But this well-intentioned initiative was naïve: 'For I have learned that it has been rejected by the Synagogue, as by the Church.' In the preface to his book on the pre-Adamites, he wrote with apparent candour: 'If someone shows me with

knowledge of cause that I have failed, in other words, that I have been wrong in the least circumstance, from the history of Genesis or any other passage in the Holy Canonic Scripture, or that I have in the least way departed from any article of the Christian Faith …', he was prepared to amend: 'For while I love myself dearly, I love the truth far more so, of which I make a singular profession.'

26 Remember that the canonical date for the Flood, fixed without ambiguity in the Bible used at this time, was 2348 BCE!

27 Martino Martini, *Histoire de la Chine*, Vol. I (C. Babin & A. Seneuze, 1692).

28 Etienne Souciet, *Observations mathématiques, astronomiques, géographiques, chronologiques et physiques, tirées des anciens livres chinois ou faites nouvellement aux Indes, à la Chine et ailleurs, par les Pères de la compagnie de Jésus* (Rollin, 1732).

29 Catherine Jami, private communication, 2008.

30 Martin Gorst, *Measuring Eternity: The Search for the Beginning of Time.*

31 We should note that one of the issues in the correct dating of the Creation was to predict the *parousia*, i.e., the second coming among men of Christ the redeemer.

32 Tony Volpe, *Science et théologie dans les débats savants de la seconde moitié du XVIIe siècle* (Brepol, 2008).

33 Virgile Pinot, *La Chine et la formation de l'esprit philosophique en France (1640–1740)* (Paul Geuthner, 1932). Downloadable at classiques.uqac.ca/classiques.

34 To quote here from Finkelstein and Silberman, *The Bible Unearthed*:

> The first question was whether Moses could really have been the author of the Five Books of Moses, since the last book, Deuteronomy, described in great detail the precise time and circumstances of Moses' own death. Other incongruities soon became apparent: the biblical text was filled with literary asides, explaining the ancient names of certain places and frequently noting that the evidences of famous biblical events were still visible 'to this day'. These factors convinced some seventeenth century scholars that the Bible's first five books, at least, had been shaped, expanded, and embellished by later, anonymous editors and revisers over the centuries.
>
> By the late eighteenth century and even more so in the nineteenth, many critical biblical scholars had begun to doubt that Moses had any hand in the writing of the Bible whatsoever; they had come to believe that the Bible was the work of later writers exclusively. These scholars pointed to what appeared to be different versions of the same stories *within* the books of the Pentateuch, suggesting that the biblical text was the work

of several recognizable hands. A careful reading of the book of Genesis, for example, revealed two conflicting genealogies of the creation (1:1–2:3 and 2:4–25), Adam's offspring (4:17–26 and 5:1–28), and two spliced and rearranged flood stories (6:5–9:17). In addition, there were dozens more doublets and sometimes even triplets of the same events in the narratives of the wanderings of the patriarchs, the Exodus from Egypt, and the giving of the Law.

Here is a quite typical reaction to *The Bible Unearthed* on the part of the contemporary Catholic church. It appeared in the *Bulletin de littérature ecclésiastique* published in 2008 by the Institut Catholique de Toulouse:

> In this respect, as J. Briend has noted, the present book shows very well the difficulty that the archaeologist has in making use of biblical texts without intending to simplify their complexity, whereas the comparison between the Bible and archaeological discoveries demands that each discipline be respected. The two in fact belong to different worlds. To maintain this does not mean that the contribution of the archaeologist is unimportant to the reading of the Bible, but it does mean acknowledging that the biblical text is the bearer of a meaning that lies beyond archaeological and historical facts alone. As the fruit of successive readings, it is the vehicle of a promise that is extended across the centuries, and in which each person may recognize himself. In this sense, it is genuinely historical.

35 Israel Finkelstein and Neil Asher Silberman, *The Bible Unearthed: Archaeology's New Vision of Ancient Israel and the Origin of Its Sacred Texts* (Free Press, 2001).

36 Shlomo Sand, *The Invention of the Jewish People* (Verso, 2009).

37 This information is taken mainly from Hermine Hartleben, *Jean-François Champollion 1790–1832, sa vie et son oeuvre* (Pygmalion, 1997), a great biography of Champollion, as well as from Jean Lacouture, *Champollion, une vie de lumières* (Grasset, 1989).

38 Bonaparte, in Egypt, does not seem to have been troubled by biblical chronology, as witness the celebrated sentence attributed to him: 'Soldiers, from the height of these pyramids, forty centuries are watching you!'

39 Baron Georges Cuvier, *A Discourse on the Revolutions of the Surface of the Globe, the Changes Thereby Produced in the Animal Kingdom* (Whittaker, Treacher & Arnot, 1829).

40 In order to avoid the cardinal's hat, Champollion cited a matrimonial situation that was not very edifying for a man of the church. The pope then simply recommended that Charles X award him the Légion d'honneur.

41 Hermine Hartleben, *Jean-François Champollion 1790–1832, sa vie et son oeuvre* (Pygmalion, 1997).

2. The Beginning of the Modern Age

1 Even in the late nineteenth century, Kelvin criticized a good number of geologists for ignoring this universality and accepting as viable theories that openly violated the principles of thermodynamics.

2 Just as Galileo had published his *Dialogue Concerning the Two Chief World Systems* in Italian, with a view to its being widely understood, so Descartes published his *Discourse on Method* in French. He justified this choice as follows in a letter to Père Vatier: 'in a book, from which I wanted even women to be able to learn something, yet that the more subtle would find sufficient material to occupy their attention.'

3 Some orthodox Jews make similar use of Psalm 90 of the Old Testament (Psalms 90:4): 'for in thy sight a thousand years are as yesterday'. Muslims use multipliers drawn from the Koran, as in sura 70, verse 4: 'The angels like the Spirit ascend towards Him in a day whose measure is fifty thousand years.' Seyyed Hossein Nasr, *An Introduction to Islamic Cosmological Doctrines* (Belknap Press of Harvard University Press, 1964), explains the arguments of the great mathematician Al-Biruni on this point.

4 Isaac Newton, *The Correspondence of Isaac Newton*, volume II (Cambridge University Press, 1960).

5 St Augustine, more than eleven hundred years earlier, had already raised this problem in *On Genesis*: 'It may seem here that time began when the heaven and the lights of heaven appeared and began their revolutions. Now, if this is true, if time began with the course of the heavenly bodies that are said to have been made on the fourth day, how could there have been days before the existence of time?'

6 Georges-Louis Leclerc Buffon, *Histoire naturelle générale et particulère, Recherches sur le refroidissement de la Terre & des Planètes*, Vol. V (Imprimerie royale, 1749a), available at www.buffon.cnrs.fr/correspondance.

7 German medical doctor and naturalist (1715–59).

8 Georges-Louis Leclerc Buffon, *Les époques de la Nature*. 1779. (New edition by Jacques Roger, including the manuscript. Editions du Muséum, 1962).

9 Georges-Louis Leclerc Buffon, *Histoire naturelle générale et particulère*, Vol. I.

10 Buffon again wrote:

> Nothing better characterizes a miracle than the impossibility of explaining its effect by natural causes; our authors have made vain efforts to account for the flood, their errors of physics on the subject of the secondary causes that they employ, prove the truth of the fact as it is reported in Holy Scripture, and show that this can only have been effected by the first cause, by the will of God ... So the universal flood has to be regarded as a supernatural means that divine Omnipotence used for the punishment of men, and not as a natural effect in which everything happened according to the laws of physics. The universal flood is thus a miracle in its cause and its effects; we see clearly from the text of holy scripture that it served solely to destroy man and animals; but, it will be said, as the universal flood was a certain fact, is it not permitted to discuss the consequences of this fact? At the right time; but you have to start by admitting that the universal flood could not have been caused by physical forces, you have to recognize it as an immediate effect of the will of the Omnipotent, you have to confine yourself to knowing simply what the sacred books teach us of it, to admit at the same time that you are not allowed to know more, and above all not mix bad physics with the purity of the holy book. Taking these precautions demanded by the respect that that we owe to the decrees of God, what is left to examine on the subject of the flood?'

11 Does not Buffon insist too much? We may feel certain doubts as to his sincerity when writes (maliciously?), again in his *Histoire naturelle*:

> and that the dove brought an olive branch; for although M. de Tournefort may claim that there are no olive trees within more than 400 leagues of Mount Ararat, and make rather poor jokes on this score (*Voyage du lévant*, vol. 2, p. 336), it is certain however that there were so in these parts at the time of the flood, since the sacred book assures us, and it is not surprising that in a space of 4,000 years the olive trees should have been destroyed in these regions and multiplied in others; it is thus wrong and against the letter of holy scripture that these authors supposed that the Earth was totally different before the Flood from how it is today, and this contradiction between their hypotheses and the sacred text, as well as their opposition to physical truths, must lead us to reject their systems, even though they do agree with certain phenomena.

12 For example, the method based on the multiplication of the human population since Noah's ark. Given that after the Flood there remained only two human beings, and assuming a given rate of growth, it is easy to calculate the time needed to obtain the present population. This method was not followed up.

13 We can mention a further technique of original dating based on variations in the magnetic field, the same field that governs compasses. By definition, these point to the magnetic north, currently close to the geographic North Pole which is defined as the intersection of Earth's axis of rotation with its surface. But this has not always been the case; over geological time, the position of the 'north' magnetic pole has varied considerably. Some 780,000 years ago it was even close to the geographic South Pole! We understand very well now, even if its mechanism remains mysterious, the chaotic history of these fluctuations, thanks to little magnets trapped by the cooling of volcanic lava. This geomagnetic calendar, when it is available for consultation, makes calculation possible.

14 Daniel Becquemont, 'L'Église anglicane face à l'évolutionisme', in *Actes du colloque sciences humaines et religion*, EHESS, Paris, September 2007.

15 Paul-Henri Thiry Holbach, 'Fossiles' (article), *Encylopédie de Diderot et D'Alembert*, 1758:

> It was already observed in the most distant antiquity that the Earth contained a very large number of marine animals; this led to the idea that it had previously been a seabed. It appears that this was the view of Xenophanes, the founder of the Eleatic sect; Herodotus observed the shells that were found in the mountains of Egypt, and suspected that the sea had retreated. That was also, according to Strabo's report, the view of Eratosethenes who lived in the time of Ptolemy Philopator [pharaoh in the third century BC] and of Ptolemy Epiphanes [immediate successor of Philopator]. The same thing was believed in Ovid's day, as in a well-known passage of his *Metamorphoses*, book 15, he wrote:
>
> > *Vidi ego, quod fuerat quondam solidissima tellus*
> > *Esse fretrum. Vidi factas ex aequore terras,*
> > *Et procul a pelage conchae jacuere marinae, &c.*
> >
> > [I have seen with my own eyes the solid earth rise up, and I have seen the sun disappear into the sea. Far from the sea, its shells are strewn on the ground.]
>
> This was also the feeling of Avicenna and the Arab scholars; but although it was so universally widespread among the ancients, it was subsequently forgotten; observations of natural history were entirely neglected among us in the centuries that ensued. When people began again to observe, the scholars to whom Peripatetic philosophy and the subtleties of scholasticism had led to adopt a very strange manner of reasoning, claimed that the shells and other

fossils foreign to the earth had been formed by a sculptural force (*vis plastica*) or by a universally widespread seeding (*seminium* and *vis seminalis*). We thus see that they regarded the marine fossil bodies as if they were tricks of nature, without paying attention to the perfect analogy that there was between these very bodies taken from within the earth, and other bodies of the sea, or belonging to the animal or the vegetable kingdom: an analogy that alone would have sufficed to undeceive them. It was felt however that there were fossil bodies to which this formation could in no way be attributed, because an organic structure was clearly observable in them: hence, for example, the opinion of several authors who regarded the fossil remains found on many parts of the earth as having belonged to the giants spoken of in holy scripture; however a little knowledge of anatomy was enough to convince them that these remains, sometimes of inordinate size, had belonged to fish or quadrupeds, not to men. These supposed sculpting forces and explanations, no matter how absurd and unintelligible they are, found and still find their champions today, among whom one may count Lister, Langius and many other naturalists who are otherwise enlightened.

16 Maillet's own text (de Maillet, 1755) does not actually give a precise date, but Richet (2009) has found this figure in other editions of this book.
17 de Maillet, 1968.
18 Dalrymple, 1991; Buffon, 1779.
19 See Richet, 2009, p. 215, which shows seventeen ages proposed between 1860 and 1909, with a range of between 3 million and 1,526 million years!
20 Here is how Buffon describes his experiments:

> Supposing, as all phenomena seem to indicate, that the earth had formerly been in a state of liquefaction caused by fire; it is shown by our experiments that if the globe were entirely composed of iron or ferric material it would only have consolidated at the centre in 4,026 years, and cooled to the point of being able to touch it without burning in 46,991 years; and that it would only have cooled to its present temperature in 100,696 years; but as the Earth, in all that is known to us, appears to be composed of vitric and calcic materials that cool in less time than ferric materials, so in order to approach as near as possible to the truth, we have to take the respective cooling times of these different materials, such as we found them by the experiments in the second Report, and establish the relationship of these to the cooling of iron. Taking into this sum only glass, clay, hard limestone, marble and ferric matter, it is found that the terrestrial globe would have consolidated at the centre in around 2,905 years, cooled to the point of

being able to touch it in about 33,911 years, and to its present temperature in about 74,047 years.

21 Buffon, *Histoire naturelle générale et particulère*; Buffon, *Les époques de la Nature*.

22 Georges-Louis Leclerc Buffon, *Lettre LXXVII au Président de Brosses*, 1760, www.buffon.cnrs.fr/correspondance.

23 Due to Fourier (1768–1830), who had thought of using it to date Earth, but had not followed this up.

24 This rate of increase holds only close to the surface, and subsequently decreases.

25 This being the ratio of Earth's average thermic conductibility to its specific heat.

26 We should mention the objection that his disciple John Perry eventually dared to formulate in 1895. See p. 35.

27 Sir William Thomson, 'On the Rigidity of the Earth'. *Royal Society of London Philosophical Transactions*, Series I, 153, 1863.

28 Sir William Thomson, 'On the Secular Cooling of the Earth'. *Philosophical Magazine* 25, January 1863.

29 Sir William Thomson, 'On the Age of the Sun's Heat'. *Macmillan's Magazine*, vol. 5, March 1862.

30 Note the caution of this incidental remark, which of course was quickly forgotten.

31 Pascal Richet, *A Natural History of Time* (University of Chicago Press, 2009).

32 This North American geologist was one of the first to establish the correlation between the concentration of carbon dioxide and global warming.

33 Thomas Crowder Chamberlin, 'On the Interior Structure, Surface Temperature, and Age of the Earth'. *Science* IX, 1899.

34 Edmond Halley (1656–1743) was famous for having predicted the return of 'his' comet by using Newton's theory.

35 Patrick Wyse Jackson, *The Chronologers' Quest: The Search for the Age of the Earth* (Cambridge University Press, 2006).

36 Thanks to laser reflectors mounted on satellites, we know today that the speed at which the Moon is moving away from Earth is 3.82 ± 0.07 cms per year. The precision is admirable!

37 The currently accepted hypothesis is that the Moon was formed less than 100 million years after Earth, as the result of a collision between Earth and a planet the size of Mars.

38 Being inversely proportional to the square of the distance.

39 The deceleration due to tidal effects is already very noticeable for the Moon,

which now turns on itself only at a speed that shows us always its same side.

40 Because of the principle of conservation of kinetic energy.

41 R. Mazumder and M. Arima, 'Tidal Rhythmites and Their Implications'. *Earth-Science Reviews* 69, 2005.

42 J. P. Vanyo and S. M. Awramik, 'Stromatolites and Earth-Sun-Moon Dynamics'. *Precambrian Research* 29, 1985. The 'clock' mechanism offered by stromatolites would be as follows.

Cyanobacteria carried out photosynthesis only in the daytime: the strata observed are thus the trace of their daily activity. Besides, according to the authors cited here, stromatolites change their angle of growth as a function of the position of the Sun, so that their growth was always exactly in this direction. From summer to winter, however, the average direction of the Sun changes, so changing with it the orientation of the respective days' strata. In one year, their centre of gravity describes a kind of sine curve, as indicated in Figure 8 (p. 215). It is only necessary therefore to be able to count the number of strata in a period of this sine curve to obtain the number of days in a year. Awramil and Vanyo showed in this way that the stromatolite *Anabaria juvensis*, from the late pre-Cambrian (850 Ma, Northern Territory, Australia) recorded years of 435 days. This means that the day was then approximately 20.1 hours; in other words, the friction of the tides has caused a reduction in Earth's speed of rotation. These observations are confirmed for later dates by corals, which 400 million years ago, for example, record years of 400 days. This is a very nice theory, and an example of the original ideas that can be found to look back across such gigantic spaces of time. That is why I mention it here, though it should be noted that it has been challenged (George E. Williams et al., 'No Heliotropism in Neoproterozoic Columnar Stromatolite Growth, Amadeus Basin, Central Australia: Geophysical Implications'. *Palaeogeography, Palaeoclimatology, Palaeoecology* 249, 2007): the columns of strata have many branches that make it hard to follow the curves, and these unusual curves could be accidental, with the result that the datings from them are erroneous; finally, the numbers of days in the year that they show contradict other estimates.

43 *Nautilus* is a small marine animal in the Pacific of a 'living fossil' kind, living in the most recently added compartment of its shell. They ascend and descend daily, using the compartment as a ballast filled with nitrogen. Each rise to the surface creates a ridge in a compartment, and in present specimens each compartment has 29 or 30 ridges. It is now widely accepted that this number corresponds to the number of days in the lunar month. Examination of fossils dated no more than 25 million years ago shows that the Moon then rotated in

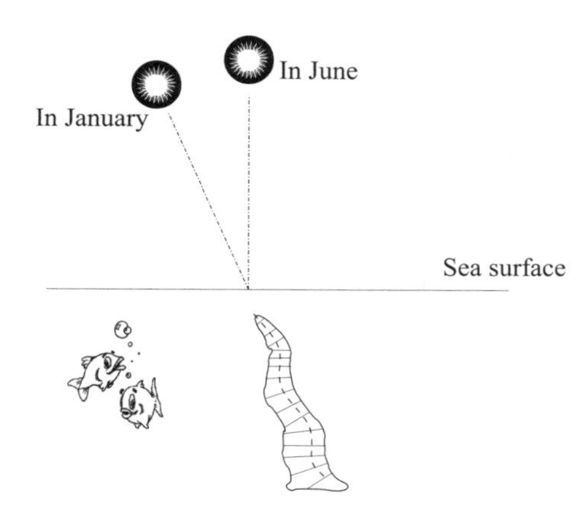

Fig. 8: Diagram showing how the annual variation in the average inclination of the Sun creates a periodic structure of strata of daily activity in the cyanobacteria. These strata are perpendicular to the average position of the Sun, indicated here in January and June. To keep the image clear, only a small fraction has been drawn.

25 days (25 ridges per compartment). 420 million years ago there were only 9 ridges; the Moon was then at only 40 per cent of its present distance from Earth, and the tides seven times stronger.

44 This explains the difference of 20 minutes between the tropical year and the sidereal year: 26,000 × 20 minutes is very close to a year.

45 See Claude Babin, *Autour du catastrophisme* (Vuibert, 2005), for a contradictory approach to the role of Cuvier's Protestantism. In a famous text (Baron Georges Cuvier, *A Discourse on the Revolutions of the Surface of the Globe, the Changes Thereby Produced in the Animal Kingdom* [Whittaker, Treacher & Arnot, 1829]) he declared: 'There is no reason to doubt but that the book of Genesis was composed by Moses himself, which would give it a still farther antiquity of five hundred years, namely, thirty-three centuries.' Further on, this great scientist, so imbued with the prejudices of his age, wrote in his quest for traces of the biblical Flood: 'The negroes, the most degraded of human beings, whose forms are the nearest to those of brutes, and whose intellect has no where expanded so greatly as to attain a regular government, nor to the least semblance of connected information, have preserved no records, no tradition. They then cannot afford us any information concerning our inquiry, although all their characters clearly show that they escaped from the great catastrophe by some other point than the Caucasian and Altaic races, from whom they were probably separated long before this catastrophe arose.'

46 James Hutton, *Theory of the Earth with Proofs and Illustrations* (William Creech, 1795). Facsimile of the Edinburgh edition (Wheldon & Wesley, 1972).

47 And which Kelvin completely accepted, despite being a fierce opponent of Lyell.

48 Stephen Jay Gould, *Time's Arrow, Time's Cycle: Myth and Metaphor in the Discovery of Geological Time* (Harvard University Press, 1988); Babin, *Autour du catastrophisme.*

49 Earth had in fact experienced four massive extinctions of flora and fauna, at least as catastrophic, before this time (c. 440, 365, 245 and 210 Ma).

50 So called by analogy with the obscuring of the Sun that would result from the explosion of several hydrogen bombs.

51 It is no longer certain today, however, that the dinosaurs were cold-blooded, like modern reptiles.

52 Charles Darwin, *On the Origin of Species by Means of Natural Selection, or the Preservation of Favoured Races in the Struggle for Life* (John Murray, 1859).

53 We should add that for Darwin these changes were always gradual: *natura non facit saltus*. This hypothesis is no longer shared today: climatic upheavals or epidemics can at certain moments very rapidly favour the multiplication of individuals bearing the 'right' genes. It is also noteworthy that T. H. Huxley, nicknamed 'Darwin's bulldog' on account of his enthusiastic propaganda, had already qualified the necessity of gradualism for defending the theory of evolution. In his review of *The Origin of Species* in the *Westminster Review*, April 1860, he wrote (p. 545): 'And Mr Darwin's position might, we think, have been even stronger than it is if he had not embarrassed himself with the aphorism, "Natura non facit saltum", which turns up so often in his pages. We believe, as we have said above, that Nature does make jumps now and then, and a recognition of the fact is of no small importance in disposing of many minor objections to the doctrine of transmutation.'

We should also add that Darwin was quite unaware of the mechanism of genes (despite himself having introduced the notion of 'gemules'), or the work of his contemporary Mendel. He accordingly found it hard to respond to the objection of dilution: just as a drop of ink disperses in a growing quantity of water, so the sample of individuals possessing good characteristics should disappear by dilution after several generations. The difference, however, is that the number of ink molecules is fixed once and for all, whereas the characteristic gene, whether dominant or recessive, is recreated each time and totally present in the individual that bears it, right down the generations.

54 In a letter to J. Croll dated 31 January 1869, Darwin wrote:

> Notwithstanding your excellent remarks on the work which can be effected within a million years, I am greatly troubled at the short duration of the world according to Sir W. Thompson, for I require for my theoretical views

a very long period *before* the Cambrian formation. If it wd not trouble you I shd like to hear what you think of Lyell's remarks on the magnetic force which comes from the sun to the earth; might not this penetrate the crust of the earth & then be converted into heat. This wd give a somewhat longer time during which the crust might have been solid; & this is the argument on which Sir W. Thompson seems chiefly to rest.

55 We should note that Ludwig Boltzmann was one of the few physicists to support Darwin; Mach and Helmholtz can also be added (Jacques Bouveresse, private communication, 2010). This was probably not by chance; Boltzmann was the founder of statistical mechanics, which introduced probabilities as a fundamental analytical tool for the laws of physics. So Boltzmann was not afraid of a theory of evolution based on random variations within species.

56 Kelvin did lose some authority by suggesting that, since life could not develop in 20 million years, and if one did not want to appeal to a *deus ex machina* – though Kelvin, as a Christian believer, could have accepted this – it would have to have been brought from outside. What seemed absurd at that time, however, seems less so today: Fred Hoyle (1915–2001) took up this idea, and today the hypothesis is not seen as so far-fetched (Pascal Picq, *Le Monde a-t-il été créé en sept jours?* [Perrin, 2009]). Francis Crick, awarded the Nobel Prize in 1962 for the discovery of DNA, has defended a similar hypothesis of panspermia.

57 Bertrand Russell, *Science and Religion* (Home University Library, 1935).

58 Daniel Becquemont, 'L'Église anglicane face à l'évolutionisme', in *Actes du colloque sciences humaines et religion*, EHESS, Paris, September 2007.

59 However, Cardinal Ratzinger, the future Pope Benedict XVI, declared in 2005 in an interesting debate with Paolo Flores d'Arcais (Joseph Ratzinger and Paolo Flores D'Arcais, *Est-ce que Dieu existe? Dialogue sur la vérité, la foi et l'athéisme* [Payot, 2005]): 'The idea, however, is that *physis*, nature, is not the product of blind chance, a blind evolution, but that behind the evolution that occurs there is a reason, and hence a moral of being itself.'

60 Yet we should avoid caricature. Bush actually sought, in his own particular way, to distance himself from the creationists, as he asserted at the same time: 'From Scripture you can gain a lot of strength and solace and learn life's lessons. That's what I believe, and I don't necessarily believe every single word is literally true.'

61 Camille Flammarion, *Astronomie populaire* (C. Marpon et E. Flammarion, 1881).

62 Steven Soter, 'Les systèmes planétaires sont-ils pleins à craquer?' *Pour la Science* 365, March 2008.

3. The Twentieth Century and Radioactivity

1 Not simply an accident, as is often implied in describing this!

2 Helium, the most abundant element in the universe after hydrogen, was first of all detected in the Sun's atmosphere (hence its name, from the Greek *helios*) by spectroscopy in 1868, then on Earth in 1895 (Ramsey) in a uranium compound, clevite.

3 This was one of the old dreams of the alchemists. The other one – to artificially bring about such transmutations – was realized by Irène and Frédéric Joliot in 1934 and earned them the Nobel Prize.

4 A. S. Eve, *Being the Life and Letters of the Rt. Hon. Lord Rutherford* (Cambridge University Press, 1939).

5 Peter Molnar et al., 'Kelvin, Perry et l'âge de la Terre'. *Pour la Science* 364, February 2008; Philip England, 'John Perry's Neglected Critique of Kelvin's Age for the Earth: A Missed Opportunity in Geodynamics'. *GSA Today* 17, January 2007.

6 John Perry, 'On the Age of the Earth'. *Nature* 11, 1895; Brian C. Shipley 'John Perry and the Age of the Earth', in *The Age of the Earth: From 4004 BC to AD 2002* (Geological Society, Special Publications, 2001).

7 Shipley, *The Age of the Earth*, is very clear on this subject.

8 Not to be confused with the period of a cyclical movement.

9 Pierre Curie showed that this period does not depend on the physico-chemical environment.

10 G. Brent Dalrymple, *The Age of the Earth* (Stanford University Press, 1991).

11 Pascal Richet (*A Natural History of Time* [University of Chicago Press, 2009]) notes none the less that many geologists, having had their fingers burned, did not immediately adopt this new chronology proposed by the physicists. Having at last accepted, somewhat reluctantly, the short timescales imposed by Kelvin, they did not see the need for the billions of years now proposed. Perhaps they also thought that if they waited a bit, the physicists would adjust their figures again …

12 The helium found in its atmosphere arises from hydrogen fusion and not, as had been believed, from the fission of a mysterious uranium that had never been detected.

13 This should not be confused with the process of alpha radioactivity, which is 20 million times more probable, and eventually leads to Pb-206.

14 Fritz Houtermans was one of the great nuclear physicists of the twentieth century, but is less well known than his accomplishments deserve. (Among other things, he established the first model of nuclear fusion, the source of

stellar energy.) Perhaps this is due to a rather chaotic personal itinerary. François Rothen, *Et pourtant, elle tourne!* (Presses polytechniques et universitaires romandes, 2004) contains a detailed description of the history of nuclear geo-chronology and a good summary of Houtermans' biography (see also Iosif B. Khriplovich, 'The Eventful Life of Fritz Houtermans', *Physics Today*, July 1992). A German Communist from 1920 on, Houtermans had to flee to England after Hitler came to power in 1933. Naturally attracted by the 'socialist fatherland', he accepted a place at Kharkhov, in the Soviet Union. Then came Stalin's great purges of 1937. Arrested and tortured, he was forced to 'confess' to having been both a Trotskyist and a German agent. In 1940, after the signing of the German-Soviet pact, he was handed over to the Gestapo along with other German anti-fascists. Once again imprisoned, this time in Berlin, he was released thanks to the intervention of Max von Laue, and could continue to work, protected by a number of his colleagues. He had the idea that the capture of a neutron by uranium-238 would produce a fissile element (which was not yet called plutonium), and helped to set Weizsäcker and Heisenberg, who were working on the project of a German atom bomb, on this track. At the same time, however, he managed to send a message via Switzerland indicating German advances to his colleagues working in the United States on the Manhattan Project of making an atom bomb. Working in Berne after the war, he conceived the isochrone method. In 1951, he published a study of the Soviet purges of the 1930s and the methods used by the NKVD to obtain 'confessions' from the detainees, under the pseudonym F. Beck, the real name of a historian who had been imprisoned with him (F. Beck and W. Godin, *Russian Purge* [Hurst & Blackett, 1951]).

15 See Appendix A.

16 Understanding this expansion of the universe is not intuitive: what is involved is a stretching of space itself, rather than an expansion of celestial bodies in a fixed space.

17 A parsec is approximately 3.3 light-years.

18 A by-product of Patterson's work on lead was his eventually successful campaign against lead additives in petrol.

19 Gérard Manhés et al., 'The Major Differentiation of the Earth at 4.45 Ga.' *Earth and Planetary Science Letters* 267, March 2008.

20 The oldest rocks found on Earth date back some 4.28 billion years. Rock datings of 4.36 billion have been made in Australia, but these are only scattered mineral grains known as zircons. They tend to prove the presence of water by this date.

21 Jeff Cuzzi, 'The First Movement'. *Nature* 448, 2007; Manhès et al., 'The Major

Differentiation of the Earth at 4.45 Ga'; Jean-Pierre Bibring, *Mars, planète bleue?* (O. Jacob, 2009).

22 This paragraph has been much assisted by Jean Duprat.

23 Jonathan O'Neil et al., 'Neodynium 142 Evidence for Haden Mafic Crust'. *Science* 321, 2008.

24 The micrograms of dust from comets that have been collected in the Antarctic with this aim in mind are by far the most expensive powder in the world.

25 O'Neil et al., 'Neodynium 142 Evidence for Haden Mafic Crust'; Yuri Amelin et al.,'Lead Isotopic Ages of Chondrules and Calcium Aluminium-Rich Inclusions'. *Science* 297, 2002.

26 Patrick N. Wyse Jackson, 'William Thomson's Determinations of the Age of the Earth', in *Kelvin: Life, Labours and Legacy* (Oxford University Press, 2008).

27 *Pour la Science*, 2004.

28 It has even been used to recalibrate another theory that is equally strong but appears to be rather less so, that of carbon-14 dating, which assumed a constant isotopic composition of Earth's atmosphere between carbon-14 and carbon-12. We know now that this has varied slightly.

29 For a discussion of the notion of scientific truth, see for example Alan Sokal, *Beyond the Hoax: Science, Philosophy and Culture* (Oxford University Press, 2008), and the texts to which he refers.

30 It is impossible to reproduce the creation of Earth, but we do now know a growing number of exoplanets (planets of other solar systems). The future appearance of *observational* data on the formation processes of other such systems cannot be ruled out. Jean-Pierre Bibring, *Mars, planète bleue?* (O. Jacob, 2009).

Part Two: Earth's Movement

1 This is not a 'royal' we: Xavier Campi and Jacques Treiner helped greatly in developing and even writing this second part, particularly with ancient astronomy.

2 In the case of an aeroplane taking off, the acceleration is far sharper, so that the movement is felt as a force pressing you back in your seat.

4. Before Copernicus

1 We might also add, though in a rather different register, the separation of science from philosophy (Jean-François Revel, *Histoire de la philosophie occidentale, de Thalès à Kant* [Nil, 1994]).
2 Claude Lévi-Strauss, *The Savage Mind* (University of Chicago Press, 1966).
3 Anton Pannekoek, *A History of Astronomy* (Dover Books, 1989).
4 A staunch defender of heliocentrism, as it happened!
5 Youssef Seddick, private communication, 2009.
6 This 'reason' would not explain why the opposite is the case in the southern hemisphere. Paradoxically, on its elliptical orbit it is during the (northern hemisphere) summer that Earth is farthest from the Sun.
7 We can note in passing that the date of the Sun's positioning in this or that constellation has significantly changed over 2,000 years, on account of the precession of the equinoxes, which raises a number of problems for astrologers.
8 The NASA web site (http://mars/jpl.nasa.gov/allabout/nightsky/nightsky04-2003animation.html) has an animation showing the retrogradation of Mars in 2003.
9 Let us start from an Earth–Jupiter conjunction. A year later, Earth has described a complete circle and Jupiter has moved along its orbit, so that the conjunction will take place rather later, at the end of a supplementary fraction of a year x. As Jupiter's angular motion is only a twelfth that of Earth, it will have turned in total an angle $(1 + x)/12$. We thus have $x = (1 + x)/12$, hence $x = 1/11$.
10 These trajectories are not closed on themselves. For this to be so, it would be necessary for the period of Earth around the Sun (one year) and the period of each planet around the Sun to stand in relationships of whole numbers.
11 Micah Ross, 'La circulation des savoirs astronomiques dans l'Antiquité'. *Pour la Science* 373, 2008.
12 Otto Neugebauer, *The Exact Sciences in Antiquity* (Princeton University Press, 1952).
13 Claudius Ptolemy, *Ptolemy's Almagest* (Princeton University Press, 1998).
14 Catherine Jami, private communication, 2008.
15 See the passage from Pinot, 1932, cited above.
16 This example of the boat moving against the shore is actually very ancient, and already found in Aristotle and Ptolemy.
17 Roger Billard, 'L'astronomie indienne, investigation des textes sanskrits et des données numériques'. *Publications de l'EFED*, LXXXIII, 1971; Charles Malamoud, private communication, 2009.

18 K. Ramasubramanian, M. D. Srinivas and M. S. Sriram, 'Modification of the Earlier Indian Planetary Theory by the Kerala Astronomers (c. 1500 AD) and the Implied Heliocentric Picture of Planetary Motion'. *Current Science* 66, 1994.

19 To the point that George Sarton can write, 'documentation concerning Egyptian and Mesopotamian science is often very precise, more precise than concerning Greek science.' *Ancient Science Through the Golden Age of Greece* (Dover Publications, 1993).

20 In a different register, Herodotus (484–425 BCE) related in his *Inquiries* a legend according to which Thales supposedly predicted the solar eclipse that took place in 585 BCE. To do this, he used Babylonian astronomical tables. This legend cannot be taken seriously, as there is no regular cycle making it possible to predict a solar eclipse visible at a particular place. With the knowledge of that time, it is unimaginable that Thales could have developed a theory for predicting these. Otto Neugebauer, *A History of Ancient Mathematical Astronomy* (Springer, 1975).

21 Carlo Rovelli, *Anaximandre de Milet ou la naissance de la pensée scientifique* (Dunod, 2007).

22 The Catholic church, in its argument with Galileo, attributed heliocentrism to Pythagoras – a paternity that is hard to confirm. It would more likely be correct to speak of the Pythagorean school.

23 The medieval church adopted this view in a certain respect, placing Paradise in the sky and Hell in the depths of Earth.

24 This wrong physics does contain elements of truth, for milieus that exert a resistance to forward motion. This is why it has remained so popular through to today. Galileo showed that, even without a force applied, a body can continue in a uniform straight-line motion, and Newton would explain that force does not create velocity but only modifies it.

25 He was, among other things, one of the first to challenge the distinction between supra- and sublunary worlds.

26 Sylvie Nony, *Abu l-Barakat al-Bagdadi: une théorie physique de la variation du mouvement au XIIe siècle, à Bagdad*. Doctoral thesis, Université Paris-Diderot Paris 7, 2010.

27 There was, however, the question of all the little irregularities not explainable in this framework, such as the irregular duration of the seasons, the gradual precession of the equinoxes, or the variation in the brightness of certain planets such as Mars or Venus.

28 Alchemy is not far away, which explains the reactions between bodies by the greater or lesser affinity of their nature.

29 The difference in timescales is clear: the accused in the great Moscow trials were rehabilitated in less than twenty-five years, whereas Galileo's situation is still not settled after four centuries.

30 To be more precise, it was the thought of Aristotle monopolized and then fossilized by the church, just as Stalin later monopolized – and then travestied – that of Marx.

31 *Simplicius, sa vie, son oeuvre, sa survie.* Actes du colloque international de Paris, 28 September–1 October 1989.

32 Sir Thomas Heath, *Aristarchus of Samos, the ancient Copernicus; a history of Greek astronomy to Aristarchus, together with Aristarchus' Treatise on the sizes and distances of the sun and moon: a new Greek text with translation and notes* (Oxford University Press, 1913).

33 Ilan Vardi, *Archimedes, the Sand Reckoner*, 2007, www.lix.polytechnique.fr/Labo/Ilan.Vardi/sand_reckoner.ps.

34 Geoffrey E. R. Lloyd, *Greek Science after Aristotle* (W. W. Norton & Co., 1973).

35 Jacques Gapaillard, *Et pourtant, elle tourne! Le mouvement de la Terre* (Seuil, 1993).

36 This was the construction of the Danish astronomer Tycho Brahe (see pp. 77–9).

37 Hipparchus showed how it is possible, instead of shifting Earth to O′, to assume that it is not the Sun that uniformly describes a circle centred on Earth at point O, but a point C, the centre of a small circle (epicycle) on which the Sun P turns with the same angular velocity. Geometrically, as shown in Figure 26, the two models are equivalent. The Sun describes a circle of the same radius, centred on O′ (Figure 26). We should note that Jean-Pierre Verdet

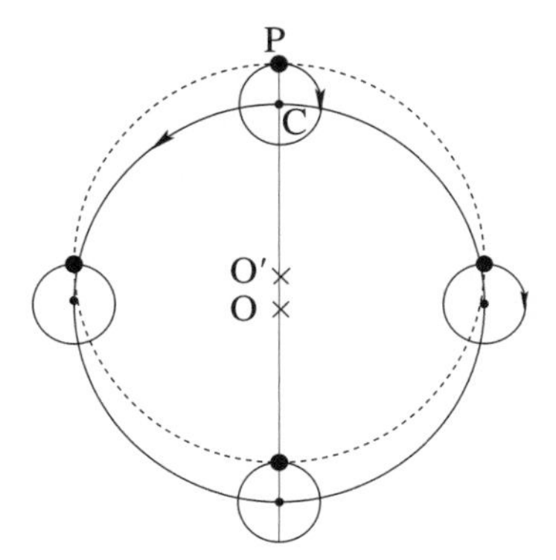

Fig. 26. Equivalence between the introduction of an eccentricity OO′ and that of the epicycle, to explain the movement of the Sun P that describes the dotted circle centred on O′.

(Verdet, *Une histoire de l'astronomie* [Seuil, 1990]) attributes the decentring of Earth already to Apollonius, which illustrates the difficulty in reconstructing Greek science, given the poverty of the archives.

38 An expression of the historian and astronomer Jean-Pierre Verdet.

39 We should note that the eccentricity of Mercury remains too strong to be reproduced by this model, which had to be further complicated by creating a variable equant for it.

40 Delambre was one of the pioneers of the metre.

41 Jean-Baptiste Joseph Delambre, *Histoire de l'astronomie ancienne* (Gabay, 2005).

42 Robert Russell Newton, *The Crime of Claudius Ptolemy* (Johns Hopkins University Press, 1977).

43 Oscar Sheynin, 'The Treatment of Observations in Early Astronomy'. *Archive for History of Exact Sciences* 46:2, 1993.

44 By Arabic astronomy I mean astronomy written in Arabic, a language that played the role that English does today. Many Jewish scholars, for example, wrote in Arabic.

45 Roshdi Rashed and Régis Morelon, *Histoire des sciences arabes* (Seuil, 1997).

46 Ahmed Djebbar, *L'âge d'or de la science arabe* (Pommier, 2005).

47 To whom we owe the word 'algorithm'.

48 The Tusi-couple consists of a large circle on which a circle of half the radius rolls internally and without slipping. A point on the circumference of the small circle then describes a diameter of the large one.

49 In this model, the Moon, for example, turns in an epicycle that itself turns in another epicycle. The reason for this complication is now understood: the Moon turns around Earth, which turns around the Sun.

50 Roshdi Rashed and Régis Morelon, *Histoire des sciences arabes*; Jean-Jacques Szczeciniarz, *Copernic et la révolution copernicienne* (Flammarion, 1998).

51 See for example Sylvie Nony, *Abu l-Barakat al-Bagdadi: une théorie physique de la variation du mouvement au XIIe siècle, à Bagdad*. Doctoral thesis, Université Paris-Diderot Paris 7, 2010.

52 Seyyed Hossein Nasr, *Science and Civilization in Islam* (Harvard University Press, 1968).

53 The main advantage of this, as we shall see, was to be kinematically exact and acceptable to the church.

5. The Construction of Heliocentrism

1 *On the Revolutions of the Heavenly Bodies.*

2 To use the train metaphor once again: in a moving carriage, you see the telegraph poles pass rapidly, whereas trees farther away from the track seem almost fixed, in proportion to their distance. Here the telegraph poles play the role of the Sun and the trees that of the fixed stars.

3 For Copernicus, the stars were already 65 times farther than for Ptolemy, but they should have been more than 300,000 times farther! Jacques Gapaillard, *Et pourtant, elle tourne! Le mouvement de la Terre* (Seuil, 1993).

4 Edward Rosen, *Three Copernican Treatises* (Dover Books, 2004).

5 Ibid.

6 Ibid.

7 Nicolas Copernicus, *Six Books on the Revolutions of the Heavenly Spheres* (Johannes Petreius, 1543). English translation and notes by Edward Rosen (Polish Scientific Publishers, 1978).

8 In actual fact, there were several such models, with more or less decentring of the spheres, and thus fewer or more epicycles.

9 Jacques Gapaillard, *Et pourtant, elle tourne! Le mouvement de la Terre* (Seuil, 1993).

10 We should recall that this was also the main criticism expressed by Arabic scientists.

11 Ana Rioja and Javier Ordóñez, *Teorías del universo. De los Pitagóricos a Galileo* (Síntesis, 2004).

12 Edward Rosen, *Three Copernican Treatises.*

13 Ibid.

14 Philippe Depondt and Guillemette de Véricourt, *Kepler: l'orbe tourmenté d'un astronome* (Éd. du Rouergue, 2007); Philippe Depondt, private communication, 2009.

15 Jean Kepler, *Astronomia Nova*. 1609. English translation: *New Astronomy* (Cambridge University Press, 1992). 'Made in Prague with the stubborn application of many years'.

16 This is in the end a procedure analogous to that of present-day physicists who wonder whether it is possible to understand the origin of interactions, the number of elementary particles, their mass and charge, with the aid of the laws of symmetry. It is the object of what is called the standard model and the quest for a grand unification. As for the question of the number of planets, it is only recently that it has been possible to begin to answer this (Soter, 2008).

17 Jean Kepler, *Mysterium Cosmographicum*, 1596. English translation: *Harmonies of the World* (Forgotten Books, 2008).

18 Ibid.
19 Particularly when we now know that there are two more planets: Uranus and Neptune. Pluto should no longer be included, as, though discovered in 1930, it was declassified as a planet in 2006.
20 Ibid.
21 Guy Israel, *Le système solaire* (Encyclopedia Universalis, 1983).
22 Arthur Koestler, *The Sleepwalkers: A History of Man's Changing Vision of the Universe* (Penguin Books, 1959).
23 An English medical doctor and physicist (1544–1603). His description of magnetism remarkably anticipated the work of Pierre Curie four hundred years later.
24 Once again, the idea that a force is needed to maintain movement is wrong, even if it is still shared today by a large portion of the so-called educated population. About Maimonides, Tony Lévy writes:

> The God whose name alone is revealed by the Scriptures, would now see His existence attested by 'science'. The proof in question rested on 'scientific' examination of the physical course of the universe: the movement of the celestial spheres requires the existence of a prime mover, and this can be called the 'physical' proof.

With a little qualification, this necessity of a 'prime mover' is found in all the medieval philosophers who sought to found faith also on reason, from Avicenna to Thomas Aquinas.
25 This should be understood as: 'Whatever is in motion must be moved by something'.
26 For Koestler, only Aristotle played a comparable role in the history of thought. We should note that both Newton and Aristotle were equally interested in the shape, age, and motion of Earth. For Aristotle these were separate questions, but for Newton they were linked.
27 This was one of Keynes' last writings.
28 The energy of big particle accelerators is needed to achieve this dream of transmutation.
29 Galileo Galilei, *Sidereus Nuncius* (Tomaso Baglioni, Venice, 1610). English translation: *Sidereus Nuncius or The Sidereal Messenger* (University of Chicago Press, 1989).
30 Newton was responsible for introducing the milling of the edges of gold coins, to prevent false coins being made by clipping good ones.
31 Isaac Newton, *Philosophiæ Naturalis Principia Mathematica*, 1687.
32 Eugène Wigner, 'The Unreasonable Effectiveness of Mathematics in the

Natural Sciences'. *Communications in Pure and Applied Mathematics* 13 (1), February 1960.

33 For recent articles in this discussion, see for example Richard Wesley Hamming,'The Unreasonable Effectiveness of Mathematics'. *The American Mathematical Monthly* 87 (2), February 1980; Jean-Louis Krivine, 'Wigner, Curry et Howard, La déraisonnable efficacité des mathématiques'. In *Exposé au colloque ARCo'04*, Sciences cognitives. Université de Compiègne, December 2004, www.pps.jussieu.fr/~krivine/articles/arco.pdf; Jeff Raskin, 'A Reply to Eugene Wigner's Paper, "The Unreasonable Effectiveness of Mathematics in the Natural Sciences" and Hamming's essay, "The Unreasonable Effectiveness of Mathematics"'. *Akadeemia* 7, 2007.

34 Bruno Latour, *Science in Action: How to Follow Scientists and Engineers through Society* (Harvard University Press, 1987).

35 Claude Allegre, *France-Soir*, 23 November 1999.

36 Connecting a function to its derivatives.

37 Pierre Simon Laplace, *A Philosophical Essay on Probabilities* [1825] (Dover, 1995).

38 That is, the very small change in initial conditions, or no matter what small perturbation in the environment.

39 Henri Poincaré, *Calcul des probabilités* (Jean Gabay, 1987).

40 If only two decimal places are used, then π is written 3.14, and $1.25 \times 1.25 = 1.56$ instead of 1.5625. It is possible to use ten or twenty decimal places instead of two, but this does not change the basic situation.

41 This is what the meteorologist Edward Lorenz observed one day in 1961 when, wanting to save time and restart a calculation on the basis of intermediate values, he manually entered printed values with three decimal places, whereas the computer, if it had restarted the calculations from the beginning, would have used at the same stage values stored in its memory with six decimal places. Instead of obtaining a second column of figures slightly different from the first, Lorenz obtained values that increasingly departed from the values in the first column, soon having no connection with these. It was from this that he concluded the *physical* instability of the system he was studying (a reduced model of the atmospheric system). This was his famous communication of 1972 to the American Academy for the Advancement of Science, entitled 'Predictability: Does the Flap of a Butterfly's Wings in Brazil Set Off a Tornado in Texas?' This date is seen as the rediscovery of what in 1975 was called chaos theory – 'rediscovery' as Poincaré was not followed up, and without mechanical means of calculation it would have been very difficult to experiment in this domain.

42 See Laskar and Gastineau, 'Existence of Collisional Trajectories of Mercury, Mars and Venus with the Earth'. *Nature* 459, June 2009, and the articles cited there.

43 The first French translation of Jean Kepler, *Astronomia Nova* (1609), did not appear until 1979.

44 'I do not propose hypotheses [i.e., speculations]'.

45 David Goodstein and Judith R. Goodstein, *Feynman's Lost Lecture* (Norton, 2000).

46 The fame of both Pannekoek and Koestler is not due just to their respective work on the history of astronomy, remarkable as this is. Both had rather tumultuous lives; they were committed Communist militants who dared to break with the Third International under Stalin. Anton Pannekoek (1873–1960), born in Holland, was known above all as a militant of the left wing of the Communist party whom Lenin polemicized against. He was a theorist of 'council communism', or the key role of workers' councils in the revolutionary process. He was also an important professional astronomer: there is an Anton Pannekoek Institute in Amsterdam. Arthur Koestler (1905–83) was born in Hungary. He took part in the Spanish Civil War and broke with the party in 1938, in the wake of the Moscow trials. His best-known work as a novelist is *Darkness at Noon*, which denounced the mechanism of 'confessions' in the course of these trials.

47 Jean-Baptiste Joseph Delambre, *Histoire de l'astronomie ancienne* (Gabay, 2005); Anton Pannekoek, *A History of Astronomy* (Dover Books, 1989); Arthur Koestler, *The Sleepwalkers: A History of Man's Changing Vision of the Universe* (Penguin Books, 1959); Alexandre Koyré, *Études d'histoire de la pensée scientifique* (Gallimard, 1973); Alexandre Koyré, *The Astronomical Revolution: Copernicus—Kepler—Borelli* (Dover, 1992); Otto Neugebauer, *The Exact Sciences in Antiquity* (Princeton University Press, 1952); Otto Neugebauer, *A History of Ancient Mathematical Astronomy* (Springer, 1975); Gérard Simon, *Sciences et histoire* (Gallimard, 2008); and Jean-Pierre Verdet, *Une histoire de l'astronomie* (Seuil, 1990).

48 See, for example, Brown (Peter Brown, *The Making of Late Antiquity* [Harvard University Press, 1978]) for late antiquity and Grant (Edward Grant, *Science and Religion, 400 B.C. to A.D. 1550: From Aristotle to Copernicus* [Johns Hopkins University Press, 2006]) for the Middle Ages.

49 Cf. S. G. Soal and Frederick Baterman, *Modern Experiments in Telepathy* (Yale University Press, 1954).

50 Philippe Depondt and Guillemette De Véricourt, *Kepler: l'orbe tourmenté d'un astronome*; David Sénéchal, 'Histoire des Sciences', in *Notes de Cours.*

Université de Sherbrooke, 2004, www.librecours.org/documents/20/2079.pdf; Jacques Blamont, *Le Chiffre et le Songe, Histoire politique de la découverte* (O. Jacob, 2005); Jacques Gapaillard, *Et pourtant, elle tourne! Le mouvement de la Terre* (Seuil, 1993).

51 Ana Rioja and Javier Ordóñez, *Teorías del universo. De los Pitagóricos a Galileo* (Síntesis, 2004).

6. Distances

1 J. L. Berggren and Nathan Sidoli, 'Aristarchus's *On the Sizes and Distances of the Sun and the Moon*: Greek and Arabic Texts', *Archive for the History of the Exact Sciences* 61, no. 3, 2007, pp. 213–54.

2 This is an example of the modern establishment of so-called experimental data, which also carry a theoretical charge.

3 Anton Pannekoek, *A History of Astronomy*, gives an eloquent picture of this first international collaboration:

> When the time came near, [Halley's] appeal met with a large response. A number of French and English astronomers journeyed to far-distant and little-known places. In 1761 the phenomenon could be seen in its entirety over Asia and the north polar regions; on the Australian islands the beginning only was visible; in western Europe and the Atlantic the end only. Pingré went to the island of Rodrigues in the Indian Ocean; Chappe d'Auteroche to Tobolsk in Siberia; Maskelyne to St Helena; Mason and Dixon to the Cape of Good Hope; Father Hell from Vienna to Vardö in Norway near the North Cape; Le Gentil to India. And in still greater number astronomers set out in 1769, when the complete phenomenon was visible over the Pacific, western America, and, of course, the North Pole; the end in eastern Asia; and the beginning in eastern America and western Europe. Chappe went to California where he died from pestilence contracted in fulfilling his task. Pingré went to San Domingo, Wales to Hudson Bay, Captain Cook with some astronomers to Tahiti; several Russian observers spread over Siberia; Hell went again to Vardö to see Venus pass over the midnight sun, whereas many European and American astronomers, as well as Mohr at Batavia, observed the phenomenon at their home observatories. The results did not come up to the high expectations … The results for the solar parallax, deduced by different computers from different combinations of observations, consequently diverged far more than Halley had optimistically expected. Moreover the geographical longitude of a number

of stations was badly known and had to be derived by the observers themselves by means of Jupiter's satellites or, in 1769, from a partial eclipse of the sun. Thus all sorts of values between 8.55″ and 8.88″ were computed and published for the solar parallax. This, however, meant enormous progress in our knowledge. Instead of by many seconds, the separate results differed by some tenths of a second only, whereas formerly the solar parallax and the distance of the sun were uncertain to ⅓ or ¼ of their value. So it may be said that the transits of Venus in the eighteenth century answered their purpose completely.

4 Steven Soter, 'Les systèmes planétaires sont-ils pleins à craquer?' *Pour la Science* 365, March 2008; J. Laskar and Gastineau, 'Existence of collisional trajectories of Mercury, Mars and Venus with the Earth'. *Nature* 459, June 2009.

5 It is tempting to make a parallel between the Titius-Bode law and Balmer's (1825–98) formula. This formula, initially established empirically, perfectly reproduces the series of emission frequencies ν observed for the hydrogen atom:

$$\nu = R\,(1 - 1/n^2),$$

where R is the Rydberg constant and n an integer greater than 1. The *subsequent* development of quantum mechanics made it possible to deduce this formula exactly from the theory. Could the Titius-Bode formula have a similar status?

7. The Battle over Heliocentrism

1 It is rather naïve to believe that, as on the subject of Democritus' (fifth century BCE) atomism, we are dealing with precursors of genius. These were largely philosophical speculations with no empirical evidence. Advances in physics came about by developing (and then combatting) the thinking of Aristotle, who was against both atomism and heliocentrism.

2 Psalm 96, verse 10, was also cited: 'Declare among the nations, "The Lord is king. He has fixed the earth firm, immovable; he will judge the peoples justly."'

3 Cardinal Bellarmine, who would later 'counsel prudence' to Galileo, presided over the trial. It was the cardinal, not the martyr, who would be canonized in 1930, then made a doctor of the church (see below, 'Galileo Rehabilitated?').

4 Arkan Simaan, 'Giordano Bruno, quatre siècles après'. *Science et pseudo-science* 28, October 2009.

5 Giordano Bruno, *Opere Complete, Oeuvres complètes* (Les Belles Lettres, 2006).

6 Bertrand Levergeois, 'De l'exemple à l'alibi'. *Europe* (no. 937), May 2007.
7 Aimé Richardt, *Saint Robert Bellarmin 1542–1621, Le défenseur de la foi* (François-Xavier de Guibert, 2004).
8 Which is not surprising, given that this was in fact a heliocentric model, but described from Earth.
9 As early as 1597, he wrote to Kepler: 'For many years I have been converted to Copernicus' doctrine.'
10 St Thomas Aquinas, *Summa Theologica*, 1920. Prima pars, question 68, article 1, www.newadvent.org/summa.
11 Actually the work of Andreas Osiander, a Protestant preacher and friend of the publisher, *against* Copernicus' wishes. Copernicus, for his part, stubbornly believed – correctly – that his description was 'true', and not simply a mathematical device.
12 The way in which La Peyrère presented Copernicus is characteristic of the caution of the time:

> TO OUR MOST HOLY FATHER POPE ALEXANDER VII
>
> It is well known, Most Holy Father, that the famous Copernicus, in establishing his Astronomic System of the motion of the Earth, did not set out to overthrow the divine inventions and observations of all the Astrologers who preceded him, nor everything that the great Ptolemy had demonstrated in so excellent a fashion. And he in no way claimed to overthrow by his hypothesis this marvellous order that we see in all celestial things. Nor did he want to confuse the rising and setting of the stars, nor trouble their direct or retrograde course; nor to destroy the appearances, theories, aspects and configurations, that the Planets form among themselves, or that they represent by their encounters with the fixed Stars. The thinking of so subtle an Astronomer was to retain the order that all the Stars keep, and to retain the demonstrations that Ptolemy had made of these, on the old hypothesis of the celestial movement. And his design was only to show that this same order, and that these same demonstrations, were, both far more convenient, and far more short, and far more clear, on the new hypothesis (albeit false) of the movement of the earth than they were on the hypothesis (which is true) of the heavenly motion.

13 Nicolas Copernicus, *De revolutionibus orbium coelestium*. Norimbergae, J. Petreium, 1543a. English translation: *Six Books on the Revolutions of the Heavenly Spheres*, 2008, www.webexhibits.org/calendars/year-text-Copernicus.html.
14 This idea, particularly propagated by Koestler (*The Sleepwalkers*), is deemed

very exaggerated by the astronomer Owen Gingerich (*The Book Nobody Read: Chasing the Revolutions of Nicolaus Copernicus* [Walker Publishing Company, 2004]), a specialist in the diffusion of Copernicus' work.

15 Martin Luther, *Table Talk*, trans. William Hazlitt (Christian Classics Ethereal Library, 2004), www.ccel.org/ccel/luther/tabletalk.html.

16 The Protestants were serious people: in 1533, Michel Servet, a precursor of the circulation of the blood, was burned alive in Geneva, then under Calvin's personal authority.

17 At least, it does not explain the two daily tides. Besides, Earth's speed of daily rotation is no more than 1.5 per cent of its speed of annual rotation.

18 Galileo Galilei, *Dialogo sopra i due massimi sistemi del mondo*. Bonaventiræ et Abrahami Elzevir, 1632. English translation: *Dialogue Concerning the Two Chief World Systems: Ptolemaic and Copernican* (Modern Library, 2001).

19 See the detailed study by Massimo Bucciantini, *Galilée et Kepler, Philosophie, cosmologie et théologie à l'époque de la contre-réforme* (Les Belles Lettres, 2008).

20 The full theory of the tides is very complex (Gillet, 1998). We now know that the Coriolis effect plays a role in it, if only marginally. This effect is actually evidence of the movement of Earth. There was therefore a (very small) grain of truth in Galileo's false explanation. Pierre Souffrin (Souffrin, 2000) offers a partial and original rehabilitation of Galileo's theory of the tides.

21 Egidio Festa, *L'erreur de Galilée* (Austral, 1995).

22 An aspect well documented by Paul Feyerabend, *Against Method* (New Left Books, 1975).

23 Identified as rings only later.

24 Egidio Festa, *L'erreur de Galilée.*

25 Jean-Yves Boriaud, *Galilée, l'Église contre la science* (Perrin, 2010).

26 Jean Kepler, *Mysterium Cosmographicum*, 1596. English translation: *Harmonies of the World* (Forgotten Books, 2008).

27 Edward Grant, *Science and Religion, 400 B.C. to A.D. 1550: From Aristotle to Copernicus.*

28 Other more directly political factors were also involved: relations between the Spanish crown, attached to a certain orthodoxy, the French monarchy confronted with the Jesuits, and the Vatican (Pietro Redondi, *Galilée hérétique* [Gallimard, 1985]).

29 This congress began in 1545 and lasted, with interruptions, for eighteen years.

30 We find in St Augustine, *The Literal Meaning of Genesis*, trans. John H. Taylor (Newman Press, 1982):

> In the Holy Books, it is always necessary to examine the revelation of eternal truths, the account of events, the prophecies, precepts and moral

counsels. On events one may wonder whether it is sufficient to take the facts in the meaning figured, whether they should still be accepted and their authenticity as historical facts supported. That there are allegories in Scripture is something that no Christian would venture to deny, needing only to remember the words of the Apostle when he says: All these things will come to them to serve them as figures.

31 This is what St Thomas Aquinas wrote (*Summa Theologica*):

> St Augustine taught that there are two rules to observe in these questions: 1) Hold indefectibly that the Holy Scripture is true. 2) When the Scripture can be explained in several manners, no one must give one of these interpretations such an absolute support that, in the case that it were established by certain reason to be false, they would have the presumption to maintain that such is the sense of the Scripture: for fear that the Holy Scripture should come to be turned to ridicule by the heathen, and the way of faith be thereby closed to them.

32 Cited by Grant, *Science and Religion, 400 B.C. to A.D. 1550: From Aristotle to Copernicus.*

33 Nicole Oresme, *Le Livre du ciel et du monde* (University of Wisconsin Press, 1968).

34 To cite Oresme:

> of the Holy Scripture that says that the sun turns et cetera: one would say that it conformed in this part to the common human way of speaking as is done in several places, as if where it is written that God repented and that he was angered and appeased and such things that are thereby not at all as the letter sounds … almost in the same way one would say that at the time of Joshua the sun stopped still or in the time of Hezekiah it turned back, and according to all appearance: but according to the truth, the earth stopped still in the time of Joshua and advanced or hastened its movement in the time of Hezekiah and in this there was no difference as for the effect that ensued. And this way seems more reasonable than the other, as will later be explained.

35 Pierre Souffrin, 'Oresme, Buridan, et le mouvement de rotation diurne de la terre ou des cieux', in *Terres médiévales* (Klincksieck, 1993), offers a detailed (and original) study of the positions of Oresme and his contemporary Buridan.

36 Galileo does not always cite his predecessors, but it was current practice at this time not to give sources, even if the authority of the ancients was commonly appealed to.

37 Averroes, *The Philosophy and Theology of Averroes. Tractacta* (Manibhai Mathurbhal Gupta, 1921).
38 Saadia Gaon, *The Book of Doctrines and Beliefs* (East and West Library, 1946).
39 Moses Maimonides, *A Guide for the Perplexed* (E. P. Dutton, 1904).
40 Benedict de Spinoza, 'A Theologico-Political Treatise', in *The Chief Works of Benedict de Spinoza* (Dover, 1951).
41 Tony Lévy, *Figures de l'infini* (Seuil, 1987).
42 The Protestants did not accept this prohibition.
43 Peter Godman, *The Saint as Censor: Robert Bellarmine between Inquisition and Index* (Brill, 2000).
44 Definition of the Arabic language is a complex question (Bruno Levallois, private communication, 2010). The language has always existed in very varied forms. To simplify, we are speaking here of standard modern or written Arabic, as taught in school and generally used in the press and on television in all Arabic-speaking countries. This differs from the spoken Arabic dialects that exist in major regional variations. It also differs from the archaic literary Arabic of the Koran.
45 Isaac Newton, *The Correspondence of Isaac Newton*, volume II (Cambridge University Press, 1960). January 1680. From the original in King's College Library, Cambridge. In reply to Letter 246.
46 Isaac Newton, 'Seven Statements on Religion', *www.newtonproject.sussex.ac.uk/view/texts/normalized/THEM00006*
47 Isaac Newton, 'Untitled treatise'. Yahuda MS 1, *www.newtonproject.sussex.ac.uk/view/texts/normalized/THEM00135*
48 Zuraya Monroy Nasr, 'La muerte de Bruno, la condena de Galileo y la prudencia de Descartes' in *Giordano Bruno: 1600–2000*, eds Laura Benitez and Jose Antonio Robles. Facultad de filosifía y letras, universidad nacional autóma de México, 2002.
49 *Pour la Science*, 2006.
50 Already in the seventeenth century, Newton (*Philosophiæ Naturalis Principia Mathematica*) wrote:

> This most beautiful System of the Sun, Planets and Comets, could only proceed from the counsel and dominion of an intelligent and powerful being. And if the fixed Stars are the centres of other like systems, these being form'd by the like wise counsel, must be all subject to the dominion of One; especially, since the light of the fixed Stars is of the same nature with the light of the Sun, and from every system light passes into all the other systems. And lest the systems of the fixed Stars should, by

their gravity, fall on each other mutually, he hath placed those Systems at immense distances one from another.

Classically, the old parable of the watch as related by the Rev. William Paley (1743–1805) is seen as the archetype of Creationist argument, or today's 'intelligent design'; if you come across a watch on your path, by far the most likely hypothesis is that its existence cannot be due to the separate manufacture and chance assembly of the various pieces of its subtle mechanism, but results from a higher purpose (William Paley, *Natural Theology; or, Evidences of the Existence and Attributes of the Deity*, 1802):

> In crossing a heath, suppose I pitched my foot against a *stone*, and were asked how the stone came to be there: I might possibly answer, that, for any thing I knew to the contrary, it had lain therefore ever; nor would it perhaps be very easy to show the absurdity of this answer. But suppose I had found a *watch* upon the ground, and it should be inquired how the watch happened to be in that place; I should hardly think of the answer which I had before given, – that, for any thing I knew, the watch might have always been there. Yet why should not this answer serve for the watch as well as for the stone? why is it not as admissible in the second case, as in the first? For this reason, and for no other, viz. that, when we come to inspect the watch, we perceive (what we could not discover in the stone) that its several parts are framed and put together for a purpose, e.g. that they are so formed and adjusted as to produce motion, and that motion so regulated as to point out the hour of the day; that, if the different parts had been differently shaped from what they are, of a different size from what they are, or placed after any other manner, or in any other order, than that in which they are placed, either no motion at all would have been carried on in the machine, or none which would have answered the use that is not served by it. To reckon up a few of the plainest of these parts, and of their offices, all tending to one result: – We see a cylindrical box containing a coiled elastic spring, which, by its endeavour to relax itself, turns round the box … We then find a series of wheels … This mechanism being observed … the inference we think is inevitable, that the watch must have had a maker … Nor … would any man in his sense think the existence of the watch, with its various machinery, accounted for, by being told that it was one out of possible combinations of material forms; that whatever he had found in the place where he found the watch, must have contained some internal configuration or other; and that this configuration might be the structure now exhibited, viz. of the works of a watch, as well as a

> different structure. Now … would it yield his inquiry more satisfaction, to be answered, that there existed in things a principle of order, which had disposed the parts of the watch into their present form and situation. He never knew a watch made by the principle of order; nor can he even form to himself an idea of what is meant by a principle of order, distinct from the intelligence of the watchmaker … A law presupposes an agent; for it is only the mode, according to which an agent proceeds: it implies a power; for it is the order, according to which that power acts.

This is the foundation of 'natural theology', at least in its contemporary meaning (Jean-Michel Maldamé, 'La théologie naturelle. Une alliance de la foi et de la raison'. *Connaître* 22–23, December 2005). And the above is also true, a fortiori, for a living being that is infinitely more complex.

51 Bertrand Russell, *Science and Religion* (Home University Library, 1935).

52 An extract from the speech by Pope Benedict XVI delivered in St Peter's Square on 6 April 2006:

> Our knowledge, which is at last making it possible to work with the energies of nature, supposes the reliable and intelligent structure of matter. Thus, we see that there is a subjective rationality and an objectified rationality in matter which coincide. Of course, no one can now prove – as is proven in an experiment, in technical laws – that both really originated in a single intelligence, but it seems to me that this unity of intelligence, behind the two intelligences, really appears in our world. And the more we can delve into the world with our intelligence, the more clearly the plan of Creation appears.
>
> In the end, to reach the definitive question I would say: God exists or he does not exist. There are only two options. Either one recognizes the priority of reason, of creative Reason that is at the beginning of all things and is the principle of all things – the priority of reason is also the priority of freedom –, or one holds the priority of the irrational, inasmuch as everything that functions on our earth and in our lives would be only accidental, marginal, an irrational result – reason would be a product of irrationality.

Yet there are today currents inspired by Catholic scientists and theologians (Pierre Valiron and Philippe Deterre, *Chercheurs en science, chercheurs de sens* [Éd. de l'Atelier, 2009]) that clearly reject this conception of God as creator-watchmaker: for example Hans Kung in Germany, John Haught in the USA or Jacques Arnould in France.

53 Georges Lemaître, 'La culture catholique et les sciences positives'. In *Actes du VIe congrès catholique des Malines*, Vol. V, 1936.

54 To qualify this, we should mention the banning of the teaching of Aristotle in 1277 by Étienne Templier, bishop of Paris, at the request of Pope John XXI. This measure, which muzzled the words of St Thomas Aquinas, who constantly based himself on Aristotle, would have a long-term effect (Tony Lévy, *Figures de l'infini* [Seuil, 1987]; Edward Grant, *Science and Religion, 400 B.C. to A.D. 1550: From Aristotle to Copernicus* [Johns Hopkins University Press, 2006]).

55 In the struggle against rationalism, here are some extracts from *Providentisimus Deus*, the encyclical of Pope Leo XIII of 18 November 1893, 'the sixteenth year of our pontificate':

> Now, we have to meet the Rationalists, true children and inheritors of the older heretics, who, trusting in their turn to their own way of thinking, have rejected even the scraps and remnants of Christian belief which had been handed down to them ... Care must be taken, then, that beginners approach the study of the Bible well prepared and furnished; otherwise, just hopes will be frustrated, or, perchance, what is worse, they will unthinkingly risk the danger of error, falling an easy prey to the sophisms and laboured erudition of the Rationalists ... and seeing that most of them are tainted with false philosophy and rationalism, it must lead to the elimination from the sacred writings of all prophecy and miracle, and of everything else that is outside the natural order ... In the second place, we have to contend against those who, making an evil use of physical science, minutely scrutinize the Sacred Book in order to detect the writers in a mistake, and to take occasion to vilify its contents. Attacks of this kind, bearing as they do on matters of sensible experience, are peculiarly dangerous to the masses, and also to the young who are beginning their literary studies; for the young, if they lose their reverence for the Holy Scripture on one or more points, are easily led to give up believing in it altogether. It need not be pointed out how the nature of science, just as it is so admirably adapted to show forth the glory of the Great Creator, provided it be taught as it should be, so if it be perversely imparted to the youthful intelligence, it may prove most fatal in destroying the principles of true philosophy and in the corruption of morality ... This is the ancient and unchanging faith of the Church, solemnly defined in the Councils of Florence and of Trent, and finally confirmed and more expressly formulated by the Council of the Vatican. These are the words of the last:

> 'The Books of the Old and New Testament, whole and entire, with all their parts, as enumerated in the decree of the same Council (Trent) and in the ancient Latin Vulgate, are to be received as sacred and canonical. And the Church holds them as sacred and canonical, not because, having been composed by human industry, they were afterwards approved by her authority; nor only because they contain revelation without error; but because, having been written under the inspiration of the Holy Ghost, they have God for their author.'

We are not very far here from a Bible that, like the Koran, was not created by human hand, but is the immediate word of God.

56 For the new utilization of the sciences, Jean-Michel Maldamé, member of the Pontifical Academy of Sciences, maintains, in a strange catalogue, or at all events one hard for a physicist to follow:

> Mathematics has constantly been more perfected, so that today topology makes it possible to account for the phenomena of life; philosophers of nature such as Edgar Morin, Prigogine and Stengers strikingly demonstrate this. If the breaks with Aristotelianism are already ancient, as they began the Renaissance, a further step has been crossed in biology in the course of the last few years. Models issuing from information science and cybernetics describe the future. Cybernetic models introduce teleonomy and renew the link with finality. The equations of thermodynamics show a natural tendency towards order and give meaning to such expressions as order through fluctuation, order arising from chaos, as interactions are applied following the rigorous laws expressed in mathematical language. The future is thus grasped by mathematicized science.

57 Despite the wealth of their empires, the lack of great scientists in Spain and Portugal, from the Renaissance through to today, is perhaps the result of an omnipresent Inquisition that long refused to give in to these pressures. There is also the sorry decline of science in the Islamic world after its golden age, well described by Nidhal Guessoum (Guessoum, *Réconcilier l'Islam et la science moderne, L'esprit d'Averroès* [Presses de la Renaissance, 2009]). And there are no traces of Jewish physicists until the late nineteenth century; if the closure of the academic world to Jews played a role, we cannot rule out the weight of a certain religious tradition.

58 Georges Minois, *L'Église et la science* (Fayard, 1991).

59 The Orthodox tradition, despite its hierarchy, has remained very conservative; but its social, cultural and geographical scope is less than that of the Catholic church, and it is also fragmented into a number of different patriarchates.

60 Jean Kepler, *Mysterium Cosmographicum*, 1596. English translation: *Harmonies of the World* (Forgotten Books, 2008).

61 Depondt and de Véricourt, 2007. According to Stauffer (Richard Stauffer, 'Calvin et Copernic'. *Revue de l'histoire des religions* 179, 1971; Richard Stauffer, 'Avant, avec, après Copernic', in *L'attitude des réformateurs à l'égard de Copernic* [A. Blanchard, 1975]; Edward Rosen, 'Calvin n'a pas lu Copernic', *Revue de l'histoire des religions* 182, 1972), professor of church history at the Strasbourg faculty of Protestant theology, the importance of Copernicus' condemnation by Calvin and Luther has been exaggerated: we have only one or two texts that attest to this. Besides, the distribution of his work was above all the work of Protestants: Osiander (1498–1552) and Rheticus (1514–74). But one could also say that Galileo's work became known thanks to Catholics: there were not many atheists who would dare to 'come out' at this time, or indeed until the mid eighteenth century.

62 Martin Luther, *Table Talk*, trans. William Hazlitt (Christian Classics Ethereal Library, 2004), www.ccel.org/ccel/luther/tabletalk.html

63 The Jewish law contained in the Pentateuch (first five books of the Bible). Gad Freudenthal, *Science in the Medieval Hebrew and Arabic Tradition* (Ashgate, 2005).

64 The compilation of the teaching of great rabbis that makes it possible to understand the Torah and bring it alive.

65 Colette Sirat, *La philosophie juive au Moyen Âge: selon les textes manuscrits et imprimés* (Éditions du Centre national de la recherche scientifique CNRS, 1983).

66 Concerning Earth's movement, however, there is nothing in the Koran quite analogous to the episode from the book of Joshua in which God stops the Sun. Nevertheless, several *hadiths* do refer to Yüsha demanding the Sun to stop its course so that he can complete the conquest of Jerusalem before the start of the Sabbath (Mardam-Bey, private communication, 2009; ISL, 2005). For example, *hadith* 3124, vol. 4, taken from the authoritative *Authentic* of Boukhari: 'O my Lord! Hold it back in its course, that it may light us. The Sun then stopped, until Allah gave victory to his prophet.' But this *hadith* never played in the Islamic world a role similar to the book of Joshua for the Christians.

The *Chanson de Roland*, dating from the late eleventh century, also tells of a miracle of God stopping the Sun, to enable Charlemagne to pursue the Moors (it was actually the Basques!) after the battle of Roncevaux:

When the King sees the light at even fade,
On the green grass dismounting as he may,
He kneels aground, to God the Lord doth pray
That the sun's course He will for him delay,
Put off the night, and still prolong the day.
An angel then, with him should reason make,
Nimbly enough appeared to him and spake:
'Charles, canter on! Light needst not thou await.
The flower of France, as God knows well, is slain;
Thou canst be avenged upon that crimeful race.'
Upon that word mounts the Emperour again.

67 Though neither Avicenna or Averroes belonged to this movement, they did share its emphasis on the importance of rationality, and the idea that the Koran was created by man. The upholders of an un-created Koran, on the other hand, considering that this was the literal word of God, did not enjoy the same margin for interpreting the text.

68 Mohamed Ali Amir-Moezzi, *Dictionnaire du Coran* (Robert Laffont, 2007); Tariq Ramadan, *Islam, the West, and the Challenge of Modernity* (Islamic Foundation, 2001).

69 It is needless to point out that the socio-political contexts that accompanied the elaboration of these texts were in each case very different.

70 There is a parallel here with alchemy. In a very interesting book, David Sénéchal ('Histoire des Sciences', in *Notes de Cours. Université de Sherbrooke*) cites Arthephius (twelfth century):

> I maintain to you in all good faith that if you wish to explain the writings of the alchemists in terms of the vulgar sense of words, you will lose yourself in the toils of a labyrinth with no exit from which you will be unable to emerge, not having to guide you the thread of Ariadne, and, despite all your efforts, this will be so much silver wasted.

We may also recall Maimonides writing: 'They [the doctors] have therefore warned you that the things mentioned are *obscure* ...'

71 See for example the books of Israel Finkelstein and Neil Asher Silberman, *The Bible Unearthed: Archaeology's New Vision of Ancient Israel and the Origin of Its Sacred Texts* (Free Press, 2001), and *David and Salomon: In Search of the Bible's Sacred Kings and the Roots of Western Tradition* (Free Press, 2006).

72 Let us mention the very original hypothesis (Pietro Redondi, *Galilée hérétique* [Gallimard, 1985]) according to which the accusation of supporting the Copernican model actually served to *protect* Galileo. In fact, Galileo was

more or less secretly a believer in atomism, a doctrine that contradicted both Aristotle and the dogma of transubstantiation (the mystery of the transformation of wine and bread in the Eucharist into the blood and flesh of Christ). According to Redondi, adherence to such a doctrine would mark someone for burning without remission. This thesis is widely questioned, resting as it does on many assumptions and few documents. A bibliography on this subject is found in Egidio Festa, *Galileo, La lotta per la scienza* (Laterza, 2007); also Francesco Beretta, *Doctrine des philosophes, doctrine des théologiens et inquisition romaine au 17e siècle: aristotélisme, héliocentrisme, atomisme*. No. 42. *Bulletin de la MHFA, Vera Doctrina: zur Begriffsgeschichte der Lehre von Augustinus bis Descartes* (Allemagne, 2006).

73 Cardinal Poupard, *L'affaire Galilée* (Éd. de Paris, 2005).

74 All the less so in that the permission to print granted by the Holy Office came with the condition that the volume containing the *Dialogo* should reproduce Galileo's condemnation and abjuration (Annibale Fantoni, 'Galilée en procès, Galilée réhabilité?' in *Problèmes historiques posés par la 'clôture' de la question galiléenne* [Saint-Augustin, Francesco Beretta édition, 2005]).

75 This is by far the most prestigious title: there are several thousand saints – no one knows exactly how many – but against this just some thirty or so doctors. Bellarmine was named venerable (that is, a candidate for beatification) as far back as 1627, but his politically delicate canonization was adjourned for a long while (Peter Godman, *The Saint as Censor: Robert Bellarmine between Inquisition and Index* [Brill, 2000]).

76 Cardinal Caientano Bisleti, *Sacra Rituum congregatione eminentissimo et reverendissimo domino cardinali Caietano Bisleti relatore Urbis et orbis concessionis tituli doctoris et extensionis ejusdem tituli ad universam ecclesiam nec non officii et missae sub ritu dupl. de comm. doctorum pontificum in honorem S. Roberti S.R.E. card. Bellarmino e societate Iesu*. Institut Catholique de Paris, 1931. Bibliothèque de Fels, cote 109 073.

77 This very pious French physicist and philosopher carried out scholarly and authoritative work on the scientific thought of the Middle Ages. He was also the author, in 1915, of a work of physics and militant patriotism entitled *La Science allemande*, which ends up:

> in order for science to be true, it is not enough for it to be rigorous [i.e., as German science is], it also has to start and end with good sense ... *Scientia germanica ancilla scientiae gallicae* [German science is a servant of French science].

78 Cardinal Eugenio Pacelli, 'Lettres apostoliques Providentissimus Deus'. Proclamation of St Robert Bellarmine as Doctor of the Church, September 1931.

79 Egidio Festa, *L'erreur de Galilée* (Austral, 1995).

80 *De Deo Uno, Tractatus Primus* (Rome, 1629), pp. 194–5.

81 Given the importance of this text and its relative obscurity, I have reproduced the original here and thank Jean-Marc Rosenfeld for his delicate translation from church Latin.

82 Hans Bieri, *Der Streit um das kopernikanische Weltsystem im 17. Jahrhundert. Freiburger Studien zur Frühen Neuzeit*, 9 (Peter Lang, 2008).

83 Here is the full text of Galileo's sentence of condemnation, dated 22 June 1633:

> Whereas you, Galileo, son of the late Vincenzo Galilei, Florentine, aged seventy years, were in the year 1615 denounced to this Holy Office for holding as true the false doctrine taught by some that the Sun is the centre of the world and immovable and that the Earth moves, and also with a diurnal motion; for having disciples to whom you taught the same doctrine; for holding correspondence with certain mathematicians of Germany concerning the same; for having printed letters, entitled 'On the Sunspots', wherein you developed the same doctrine as true; and for replying to the objections from the Holy Scriptures, which from time to time were urged against it, by glossing the said Scriptures according to your own meaning; and whereas there was thereupon produced the copy of a document in the form of a letter, purporting to be written by you to one formerly your disciple, and in this divers propositions are set forth, following the position of Copernicus, which are contrary to the true sense and authority of Holy Scripture:
>
> This Holy Tribunal being therefore of intention to proceed against the disorder and mischief thence resulting, which went on increasing to the prejudice of the Holy Faith, by command of His Holiness and of the Most Eminent Lords Cardinals of this supreme and universal Inquisition, the two propositions of the stability of the Sun and the motion of the Earth were by the theological Qualifiers qualified as follows:
>
> The proposition that the Sun is the centre of the world and does not move from its place is absurd and false philosophically and formally heretical, because it is expressly contrary to Holy Scripture.
>
> The proposition that the Earth is not the centre of the world and immovable but that it moves, and also with a diurnal motion, is equally absurd and false philosophically and theologically considered at least erroneous in faith.
>
> But whereas it was desired at that time to deal leniently with you, it was decreed at the Holy Congregation held before His Holiness on the twenty-fifth of February, 1616, that his Eminence the Lord Cardinal Bellarmine

should order you to abandon altogether the said false doctrine and, in the event of your refusal, that an injunction should be imposed upon you by the Commissary of the Holy Office to give up the said doctrine and not to teach it to others, not to defend it, nor even to discuss it; and your failing your acquiescence in this injunction, that you should be imprisoned. In execution of this decree, on the following day at the palace of and in the presence of the Cardinal Bellarmine, after being gently admonished by the said Lord Cardinal, the command was enjoined upon you by the Father Commissary of the Holy Office of that time, before a notary and witnesses, that you were altogether to abandon the said false opinion and not in the future to hold or defend or teach it in any way whatsoever, neither verbally nor in writing; and upon your promising to obey, you were dismissed.

And in order that a doctrine so pernicious might be wholly rooted out and not insinuate itself further to the grave prejudice of Catholic truth, a decree was issued by the Holy Congregation of the Index prohibiting the books which treat of this doctrine and declaring the doctrine itself to be false and wholly contrary to the sacred and divine Scripture.

And whereas a book appeared here recently, printed last year at Florence, the title of which shows that you were the author, this title being: 'Dialogue of Galileo Galilei on the Great World System'; and whereas the Holy Congregation was afterward informed that through the publication of said book the false opinion of the motion of the Earth and the stability of the Sun was daily gaining ground, the said book was taken into careful consideration, and it there was discovered a patent violation of the aforesaid injunction that had been imposed upon you, for in this book you have defended the said opinion previously condemned and to your face declared to be so, although in the said book you strive by various devices to produce the impression that you leave it undecided, and in express terms as probable: which, however, is a most grievous error, as an opinion can in no wise be probable which has been declared and defined to be contrary to divine Scripture.

Therefore by our order you were cited before this Holy Office, where, being examined upon your oath, you acknowledged the book to be written and published by you. You confessed that you began to write the said book about ten or twelve years ago, after the command had been imposed upon you as above; that you requested license to print it without, however, intimating to those who granted you this license that you had been commanded not to hold, defend, or teach the doctrine in question in any way whatever.

You likewise confessed that the writing of the said book is in many places drawn up in such a form that the reader might fancy that the arguments brought forward on the false side are calculated by their cogency to compel conviction rather than to be easy of refutation, excusing yourself for having fallen into an error, as you alleged, so foreign to your intention by the fact that you had written in dialogue and by the natural complacency that every man feels in regard to his own subtleties and in showing himself more clever than the generality of men in devising, even on behalf of false propositions, ingenious and plausible arguments.

And a suitable term having been assigned to you to prepare your defence, you produced a certificate in the handwriting of his Eminence the Lord Cardinal Bellarmine, procured by you, as you asserted, in order to defend yourself against the calumnies of your enemies, who charged that you had abjured and had not been punished but only that the declaration made by His Holiness and published by the Holy Congregation of the Index has been announced to you, wherein it is declared that the doctrine of the motion of the Earth and the stability of the Sun is contrary to Holy Scriptures and therefore cannot be defended or held. And, as in this certificate there is no mention of the two articles of the injunction, namely, the order not 'to teach' and 'in any way', you represented that we ought to believe that in the course of fourteen or sixteen years you had lost all memory of them and that this was why you said nothing of the injunction when you requested permission to print your book. And all this you urged not by way of excuse for your error but that it might be set down to a vainglorious ambition rather than to malice. But his certificate produced by you in your defence has only aggravated your delinquency, since, although it is there stated that said opinion is contrary to Holy Scripture, you have nevertheless dared to discuss and defend it and to argue its probability; nor does the license artfully and cunningly extorted by you avail you anything, since you did not notify the command imposed upon you.

And whereas it appeared to us that you had not stated the full truth with regard to your intention, we thought it necessary to subject you to a rigorous examination at which (without prejudice, however, to the matters confessed by you and set forth as above with regard to your said intention) you answered like a good Catholic. Therefore, having seen and maturely considered the merits of this your case, together with your confessions and excuses above-mentioned, and all that ought justly to be seen and considered, we have arrived at the underwritten final sentence against you.

Invoking, therefore, the most holy name of our Lord Jesus Christ and of His most glorious Mother, ever Virgin Mary, by this our final sentence, which sitting in judgement, with the counsel and advice of the Reverend Masters of sacred theology and Doctors of both Laws, our assessors, we deliver in these writings, in the cause and causes at present before us between the Magnificent Carlo Sinceri, Doctor of both Laws, Proctor Fiscal of this Holy Office, of the one part, and you Galileo Galilei, the defendant, here present, examined, tried, and confessed as shown above, of the other part –

We say, pronounce, sentence and declare that you, the said Galileo, by reason of the matters adduced in trial, and by you confessed as above, have rendered yourself in the judgement of this Holy Office vehemently suspected of heresy, namely, of having believed and held the doctrine – which is false and contrary to the sacred and divine Scriptures – that the Sun is the centre of the world and does not move from east to west and that the Earth moves and is not the centre of the world; and that an opinion may be held and defended as probable after it has been declared and defined to be contrary to the Holy Scripture; and that consequently you have incurred all the censures and penalties imposed and promulgated in the sacred canons and other constitutions, general and particular, against such delinquents. From which we are content that you be absolved, provided that, first, with a sincere heart and unfeigned faith, you abjure, curse, and detest before us the aforesaid errors and heresies and every other error and heresy contrary to the Catholic and Apostolic Roman Church in the form to be prescribed by us for you.

And in order that this your grave and pernicious error and transgression may not remain altogether unpunished and that you may be more cautious in the future and an example to others that they may abstain from similar delinquencies, we ordain that the book of the 'Dialogues of Galileo Galilei' be prohibited by public edict.

We condemn you to the formal prison of this Holy Office during our pleasure, and by way of salutary penance we enjoin that for three years to come you repeat once a week the seven penitential Psalms. Reserving to ourselves liberty to moderate, commute or take off, in whole or in part, the aforesaid penalties and penance.

And so we say, pronounce, sentence, declare, ordain, and reserve in this and in any other better way and form which we can and may rightfully employ.

F. Cardinal d'Ascoli

B. Cardinal Gessi
G. Cardinal Bentivoglio
F. Cardinal Versopi
Fr. D. Cardinal of Cremona
M. Cardinal Ginetti
Fr. Ant. Cardinal of S. Onofrio

84 Many secular writers, such as Jacques Blamont (*Le Chiffre et le Songe, Histoire politique de la découverte*) and Arthur Koestler (*The Sleepwalkers*), have echoed this trope in an uncritical fashion. Galileo certainly had a high idea of his own worth: Was he wrong? Without his pride, would he have been Galileo? In a recent book, Giraud-Ruby (2010) goes still further in defending the objectivity of the Holy Office. For a more detailed presentation of this question, the reader may consult Annibale Fantoni, 'Galilée en procès, Galilée rehabilité?' in *Problèmes historiques posés par la 'clôture' de la question galiléenne* (Saint-Augustin, Francesco Beretta édition, 2005).

85 This is a kind of de facto infallibility, though not the same as the *ex cathedra* infallibility in matters of faith and morals as defined by Pope Pius IX at the Vatican I council of 1870.

86 Georges Minois, *L'Église et la science* (Fayard, 1991).

87 In fact, as we shall explain below, there is a very small eastward drift, due to the Coriolis force. Galileo had discovered the principle of inertia, but he wrongly applied this to circular motion.

88 Robert H. Romer, 'Foucault, Reich, and the Mines of Freiberg'. *American Journal of Physics* 51, 1983.

89 With the reservations mentioned above (p. 132). It was not until 1835 that the works of Copernicus and Galileo were definitively removed from the Index.

90 The Danish astronomer Ole Christensen Rømer (1644–1710) had shown that light has a finite speed of propagation estimated at 212,000 km/s, an underestimate of about 30 per cent. It is interesting to note that this estimate was made on the assumption that Jupiter's moon Io could not violate Kepler's laws. In measuring its period of revolution around Jupiter, however (some 42.5 hours), variations of plus or minus 10 minutes were observed. This period was measured as the interval of time between two eclipses of the moon by Jupiter. Rømer understood that the different measurements of this period corresponded to different distances of Earth from Jupiter, and thus different times that it took the light to reach us.

91 This is the angle that a matchstick would make at a distance of 20 kilometres.

92 Christian Doppler (1803–53) was an Austrian physicist.

93 Hippolyte Fizeau (1819–96) was a French physicist.

94 This happens to coincide here with the speed of the swimmer in relation to the medium of propagation: water. But what matters is the speed in relation to the source.

95 Given the simplicity of Newton's model, this value was remarkably close to that of 1/298 accepted today.

96 It is clear that the right starting point for a theory is not necessarily 'common sense' or even repeated experiment. See Duhem, in the quote on p. 152.

97 The first study of this curve was made variously by Galileo, Mersenne and Roberval.

98 The experiment was already suggested by Newton and tried out by Hooke, but with a height of only 9 metres, which was insufficient.

99 Meaning that motion has a meaning only in relation to a frame of reference. (H.K.)

100 See the start of this chapter. (H.K.)

101 Henri Poincaré, *The Value of Science* (Modern Library, 2001).

8. Why Truth Matters

1 Étienne Klein, 'De la relativité du relativisme'. *Agenda de la pensée contemporaine* 16, 2010.

2 These naturally include propositions not defined in scientific terms, of the kind 'God is love', the search for 'better ways for the salvation of the soul' or the 'true sense of life'. But it is best to avoid needless prophetic limitations: see the remarks on p. 154 on the impossibility of knowing the age of Earth, or the chemical composition and temperature of stars, etc.

3 Besides, to paraphrase Bertrand Russell's paradox, is the assertion 'there is no absolute truth' an absolute truth?

4 Jacques Bouveresse, *Peut-on ne pas croire?* (Agone, 2007).

5 A 'theory' born in the early nineteenth century that associates psychological or even social characteristics with the shape of the skull. It was used for a short period in criminology.

6 Alfred Wegener (*Die Entstehung der Kontinente und Ozeane* [Vieweg, 1915]), the discoverer of plate tectonics (continental drifting), made the following parallel between the work of the scientist and that of the judge:

> In order to reveal the former states of the globe, all the sciences concerned with the problems of the Earth must make their contribution, and it is only by combining all the indications they provide that the truth can be obtained; but this idea seems not always to be sufficiently widespread

among researchers … What is certain is that at any given time, the Earth can only have had one single surface, on which it does not offer us direct information. We face the Earth like a judge facing an accused who refuses any response, and our task is to discover the truth with the aid of assumptions. All the evidence we can provide presents the deceptive character of assumptions. What response would we give to the judge who reached his conclusion by using only a part of the indications available to him? It is only by bringing together the findings of all the sciences that relate to the study of the globe that we may hope to obtain the 'truth', i.e., the image that best systematizes the totality of known facts, and that can claim as a consequence to be the most probable. And even in this case, we must expect it to be modified, at any moment, by a completely new discovery, whatever the science that has made this possible.

7 We should remember, however, that it was not scientific conservatism that opposed Galileo, but that of the church of the time. The scientists, for their part, were rapidly won over.

8 The hypothesis of political assassination has been proposed. Galois was a militant republican, and a member of the Société des Amis du Peuple, led by Raspail.

9 Any new theory is submitted to a ferocious internal criticism that may delay its acceptance: the theory of continental drift proposed by Wegener took half a century to win the day, as he was unable to provide a convincing explanation.

10 The twin who travelled at a speed close to that of light would return having aged less than his sibling who remained on Earth.

11 In June 1988, the periodical *Nature* published an article in which a Dr Benveniste maintained that a solution so dilute that it no longer contained a single molecule of the diluted substance would preserve despite this a biological activity, thus suggesting a 'memory of water'.

12 In March 1999, two physicists, Fleischmann and Pons, believed they detected an abnormal release of heat in the course of electrolysis, which they interpreted as indicating nuclear fusion at an ordinary temperature.

13 Though received with a scepticism proportionate to the theoretical revolution it would have inspired, cold fusion was the subject of intensive study in major laboratories. The attitude of scientists was in fact determined by Pascal's celebrated wager: a proposition that has a very low probability of being true, but a tremendous possible gain (intellectual and, we should not forget, material), is worth the trouble to test. A similar attitude was taken towards the memory of water.

14 For a long time, units of human length were still used, such as inches, arms,

yards, etc. Galileo even measured time by his pulse! Whole generations grew up accustomed to natural standards of measurement: time determined by the day or the year, distance by the standard metre, weight by the mass of platinum held at the Breteuil pavilion. These intuitive representations are now a thing of the past: time is determined today by atomic frequencies, and length by the distance that light travels in a second, which is set by convention at 299,792,458 metres. The mass of platinum remains, but not for much longer.

15 We might add, as the physicist P. W. Anderson wrote in lapidary style: 'More is different.' For gigantic collections of atoms, therefore, classical statistical mechanics has to be used, which is not a new physics as such, but a way of using probability theories to obtain certain results.

16 A term preferred here to the notion of *paradigm change* (Thomas Samuel Kuhn, *The Structure of Scientific Revolutions* [University of Chicago Press, 1962]), which risks concealing the continuities. One could use good arguments to maintain, for example, that Copernicus' general philosophy was that of Aristotle.

17 In Latin languages at least, the word 'order' means both regularity and command.

18 No one has yet ventured to speak of planetary Darwinism to account for the survival of only certain orbits following the initial chaos.

19 Steven Soter, 'Les systèmes planétaires sont-ils pleins à craquer?'

20 The English word 'serendipity', which has no French equivalent, is 'the faculty of making happy and unexpected discoveries by accident'.

21 On this subject, the famous philosopher Auguste Comte wrote in 1835 (*Cours de philosophie positive* [J. B. Baillière et fils, 1864], Vol. II, 19e leçon, p. 6): 'We conceive the possibility of determining their forms, their distances, their sizes and their movements; whereas we shall never be able to study by any means their chemical composition or their mineralogical structure.' But the prophet was belied: in 1865, two German scientists, Robert Bunsen and Gustav Kirchoff, analysed the light of the Sun for the first time and used spectroscopy to reveal its chemical composition. August Comte drew similar conclusions on the impossibility of determining the Sun's temperature, which were likewise later belied, in this case by the theory of black body radiation (Wien's law, in 1896).

22 See, among other sources, William Shea, *Galileo's Intellectual Revolution: Middle Period, 1610–1632* (Science History Publications, 1977).

23 Jacques Blamont, *Le Chiffre et le Songe. Histoire politique de la découverte* (O. Jacob, 2005).

24 As he is well aware, since despite this claim, he devotes almost two hundred pages just to the battle of ideas between Galileo and the Vatican!

25 Alan Sokal and Jean Bricmont, *Intellectual Impostures* (Profile Books, 1998); Alan Sokal, *Beyond the Hoax: Science, Philosophy and Culture* (Oxford University Press, 2008).

26 There exists, for example, a Société française de physique, and not, fortunately, a Société de physique française. When an adjective (Aryan, German, Jewish, French, proletarian, etc.) is added to science, what is being referred to is often something quite different.

27 Debate about GMOs often reveals the confusion on this subject. To be for or against GMOs is a false alternative: it is perfectly legitimate to defend the interest of research in this field, while still being opposed to its exploitation by Monsanto.

28 As witness the name index in the present volume. The only *four women* cited are the marquise du Chatelêt for the translation and exposition of Newton's *Principia* (particularly for Voltaire), Marie Curie, Irène Joliot-Curie and Isabelle Stengers. Is this sexism on the author's part, the sexism of the societies described, or a combination of both?

29 See, for example, the quote from Descartes on women's intelligence, p. 208, note 2.

30 For readers unfamiliar with the philosophy of science, we can say that among the large number of modern authors, it is possible to distinguish *grosso modo* the upholders of a certain scientific relativism, or even a certain relativism period, depending on the moment at hand. Paul Feyerabend (*Against Method* [New Left Books, 1975]) believes that the application of so-called scientific criteria has a sterilizing effect and favours conservatism; he appeals therefore to a methodological anarchism. Kuhn (*The Structure of Scientific Revolutions*) considers that the value of scientific results can only be assessed in the sociocultural environment of their time, i.e., the paradigm, which changes suddenly in the course of development. Sometimes paradigms are deemed mutually 'incommensurable' – something that is far more debatable. Latour (*Science in Action*), for his part, insists that science is a social construction in which experience intervenes only as a supplementary argument in establishing the balance of forces, which alone is responsible for the victory of one theory over another. Even if we are generally opposed to this kind of philosophy, as the reader will easily have guessed, I do believe that some of these provocative views can be stimulating, and help to avoid a naïve view of scientific progress. A good discussion of scientific method as challenged by (post-)modernism and religion is well presented by Russell (*Science and Religion*), Jacques

Bouveresse (*Prodiges et vertiges de l'analogie. De l'abus des belles-lettres dans la pensée, Peut-on ne pas croire?*), Paul Boghossian (*Fear of Knowledge: Against Relativism and Constructivism* [Oxford University Press, 2006]), Alan Sokal (*Beyond the Hoax: Science, Philosophy and Culture* [Oxford University Press, 2008]), Alan Sokal and Jean Bricmont (*Intellectual Impostures*), among others.

31 Paul Veyne, *Did the Greeks Believe in Their Myths?* (University of Chicago Press, 1988).

32 Richard Dawkins, *The Blind Watchmaker* (W. W. Norton & Company, 1986).

33 This is the famously provocative formula of Feyerabend: 'anything goes'. See on p. 134 above, for example, how Cardinal Ratzinger sought to use this philosopher to justify the trial of Galileo.

Afterword

1 Marcel Proust, *Remembrance of Things Past: 1* (Penguin Books, 1981), p. 162.

2 Paul A. Boghossian, 'What the Sokal Hoax Ought to Teach Us', in Noretta Koertge (ed.), *A House Built on Sand: Exposing Postmodern Myths About Science* (Oxford University Press, 1998), pp. 26–7.

3 Bruno Latour, *Science in Action: How to Follow Scientists and Engineers through Society* (Harvard University Press, 1987), p. 99.

4 Alan D. Sokal, 'What the *Social Text Affair* Does and Does Not Prove', in *A House Built on Sand: Exposing Postmodernist Myths About Science*, ed. Noretta Koertge (Oxford University Press, 1998), p. 13.

5 Bruno Latour, *Science in Action*, p. 25.

6 Ibid., p. 104.

7 Ibid., p. 108.

8 Ibid., p. 99.

9 Larry Laudan, *Science and Relativism* (University of Chicago Press, 1990), p. x; cited by Sokal, 'What the *Social Text Affair* Does and Does Not Prove', p. 9.

10 Bruno Latour, *Science in Action*, pp. 30–31.

11 Paul A. Boghossian, 'What the Sokal Hoax Ought to Teach Us', p. 23.

12 Above, p. xiii.

13 Above, p. 157.

14 Noretta Koertge (ed.), *A House Built on Sand*, p. 7.

Appendices

1 Which can be assumed to be true locally.

2 This choice of zero is arbitrary: the only relevant quantity is the difference between the surface temperature and the temperature at the center. Indeed, Eq. B.1 is invariant under translation both in time and temperature. Imposing a surface temperature constant in time is a choice imposed by the fact that life appeared on Earth a long time ago.

3 In fact, this ratio has slightly evolved. Dendrochronological data are used to calibrate the results. The correction is of a few hundred years for the last 5,000 years.

4 This meteorite fell near a small village named Allende in the north of Mexico.

5 The curve is called a trochoïd.

Acknowledgements

1 Unfortunately, the commonplace character of an opinion gives no greater immunity.

Glossary

Aphelion The point on Earth's trajectory farthest from the Sun.

Astronomical Unit (AU) The average Earth–Sun distance, or about 150 million kilometres.

Conjunction (planetary) When the Sun, Earth and a planet are aligned in this order.

Creationism The doctrine that takes literally the stories of the Bible and/or Koran. It is diametrically opposed to the Darwinian theory of evolution, and more generally to all scientific discoveries about the formation of Earth.

Deferent In the geocentric system, a circle centred more or less on Earth, and travelled at constant speed by the centre of the epicycle of the planet in question.

Diurnal Daily (from Latin *dies*), or with the duration of a day. For example, Earth's rotation on itself in (close to) twenty-four hours is called its diurnal motion.

Eccentricity Characterizes the deformation of the ellipse. If a and b are its semi-axes, then c = root $a^2 - b^2$ and the eccentricity is given by $e = c/a$.

Ecliptic In the heliocentric system, this is the plane of Earth's orbit around the Sun. It is also roughly that of the other planets. In the geocentric system, it is the plane of the Sun's orbit as well as those of the planets.

Ellipse Etymologically, an imperfect circle. A useful image can be given from gardening technique: two posts are set in the ground with a cord looped around them. If the cord is stretched out, then walking along with it will trace an ellipse with the posts as its foci. If the two foci coincide, we have the particular case of the circle. The eccentricity e of the circle is zero. The other extreme case is that of the segment of straight line joining the two foci, when $e = 1$.

Epicycle In the geocentric system, this is the circle travelled uniformly by a planet; the centre of the epicycle itself turns on the deferent.

Equant Distance from the centre of the deferent to Earth. This (small) correction had to be added by Ptolemy to the purely geocentric system; it accounts ad hoc for the elliptical nature of orbits, which was later revealed by Kepler.

Equinox One of the two moments in the year, at a six-monthly interval, when the duration of the night is the same as that of the day. The vernal equinox, in March, marks the beginning of spring in the northern hemisphere; the autumnal equinox, in September, that of autumn.

Geocentrism Doctrine that places Earth at the centre of the solar system and the universe.

Gnomon An object (stick or obelisk) planted vertically. The measurement of the shadow cast by the Sun makes it possible to note the Sun's height. This is the principle of the solar sextant.

Heliocentrism Theory that considers that Earth and the planets turn around the Sun. It is opposed to the geocentrism of antiquity.

Intelligent Design The idea that the organization of the world, and particularly of living creatures, is so complex and subtle that it cannot be totally the result of chance. A structured project would have been needed to achieve it, in other words, a certain already-existing 'intelligence'. Its champions often refrain from invoking a God and accept modern scientific findings, but they challenge Darwinism. This is a more presentable version of creationism.

Isotopes Nuclei of an element that differ in their number of neutrons despite having the same number of protons and therefore the same chemical properties.

Kepler's laws There are three of these:

1. The planets each describe an elliptical orbit with the Sun as one of its foci.
2. The areas swept in equal times by the segment of straight line joining the Sun to a particular planet are equal.
3. The square of the period of revolution of a planet is proportional to the cube of its average distance from the Sun.

Newton would subsequently deduce these mathematically from the theory of gravitation and the laws of dynamics.

Latitude The angle formed by the vertical at a point on Earth and the equatorial plane. Used together with longitude to determine position on Earth's surface.

Light-year The distance travelled in a year by light in empty space. It corresponds to some 10,000 billion (10^{13}) kilometres. The Moon is roughly 1 light-second from Earth, and the Sun 8 light-minutes. Proxima Centauri, the nearest star, is 4.3 light-years away. The Andromeda galaxy is 2.2 million light-years distant, and the most distant galaxies observed (in 2010) are 12 billion light-years.

Longitude The angle formed by the planes determined by the meridian at the point in question and that of the point of reference (generally Greenwich). Used together with latitude to determine position on Earth's surface.

Lunation Time separating two identical phases of the Moon as seen from Earth. This is an average of 29.5 days (between 29 days, 6 hours and 29 days, 20 hours).

Meridian

- Terrestrial. The half of a great circle on the globe that passes through the poles and the point in question. This is characterized by its longitude. The Greenwich meridian is used as origin.
- Celestial. The half of a great circle on the celestial sphere that passes through the poles and the heavenly body in question. It is characterized by its *right ascension*.

Opposition When Earth, the Sun and a planet are aligned, in that order.

Parallel The circle joining points on Earth with the same latitude. These parallels are perpendicular to the meridians.

Parsec The distance at which an AU (astronomical unit) is seen at an angle of 1 second of arc. This corresponds to 3.3 million light-years.

Perihelion The point on Earth's trajectory when it is closest to the Sun.

Period of a planet

- Sidereal. The duration of the planet's trajectory around its orbit.
- Synodic. The time separating two oppositions (or two conjunctions). This period depends on both the sidereal period and the movement of Earth. It is what is directly observable on the basis of terrestrial measurements.

Period of radioactivity The time needed for a radioactive element, *whatever its age*, to lose half of its activity. The period of carbon-12, for example, is 5,730 years, that of uranium-238 is 4.5 million years. The former can be used to date historical events, and the latter cosmological ones.

Planet A body orbiting around a star and not emitting its own light. In the case of the solar system, a distinction is made between:

- the inner planets, orbiting between the Sun and Earth, i.e., Mercury and Venus;
- the outer planets, orbiting beyond Earth, i.e., Mars, Jupiter, Saturn, Uranus and Neptune.

In the case of a star other than the Sun, the term 'exoplanet' is used.

Precession of the equinoxes Change in the appearance of the sky due to the slow rotation of Earth's polar axis (see p. 23).

Quadrature When Earth, the Sun and a planet make a right angle.

Retrogradation Seen from Earth, the planets shift eastward in relation to the stars. At certain moments, however, this shift temporarily changes direction, a phenomenon known as retrogradation. The heliocentric description gives a natural explanation of this.

Solstices Dates when the day is longest (summer solstice in the northern hemisphere) or shortest (winter solstice in the northern hemisphere).

Spectroscopy This makes it possible, by studying the spectrum of the light that a body such as a star emits, to determine its chemical composition. This spectrum contains bands or stripes of absorption (dark) and emission (light) characteristic of the chemical elements.

Tide The tides are due to the fact that the gravitational forces exerted by the Moon (and to a small degree the Sun) at different places on Earth vary, being dependent on the distance of the point from the Moon and Sun respectively. These differences in strength create tensions. The most familiar are those moving the masses of water in the oceans. Tides also affect the solid parts of Earth, but the effects of these are harder to show.

Tropics Parallels situated at latitudes 23° 27′ north (Cancer) and 23° 27′ south (Capricorn), where the Sun is overhead at the June and December solstices respectively.

Year

- Tropical. This is the perceptible interval separating two consecutive spring equinoxes. It is approximately 365 days, 5 hours and 48 minutes. It is the 'customary' year based on the seasons.
- Sidereal. The interval separating, at a given point on Earth, two identical positions of the Sun in relation to the fixed stars. Essentially owing to the precession of the equinoxes, the sidereal year is (currently) 20 minutes and 24 seconds longer than the tropical year.

Zenith Point of intersection with the celestial sphere of a vertical at a point on Earth's surface. Owing to the inclination of the polar axis from the plane of the ecliptic, the Sun can only reach the zenith between the tropics of Cancer and Capricorn.

Zodiac Seen from Earth, this is the region of the celestial vault against which the Sun, Moon and planets seem to move. The region is traditionally divided into twelve parts named after constellations (which astrologers call 'signs').

Bibliography

Le Catéchisme de l'Église catholique, 1999. Pocket. § 881 and 882.

Le temps des datations, January 2004. Special member of *Pour la Science*.

Encyclopédie de l'Islam. Brill, 2005. Vol. XI, page 381.

Leibniz: le penseur universel, August 2006. Special number of *Pour la Science*.

Création contre Évolution, no. 26–27, 2007. Association Foi et Culture Scientifique.

Le temps est-il une illusion?, no. 397, November 2010. Special number of *Pour la Science*.

Simplicius, sa vie, son oeuvre, sa survie, 1989. Actes du colloque international de Paris (28 September–1 October).

Claude ALLÈGRE: *France-Soir*, 23 November 1999.

Yuri AMELIN et al.: 'Lead Isotopic Ages of Chondrules and Calcium Aluminium-Rich Inclusions'. *Science* 297, 2002.

Mohamed Ali AMIR-MOEZZI: *Dictionnaire du Coran*. Robert Laffont, 2007.

St THOMAS AQUINAS: *Summa Theologica*, 1920. Prima pars, question 68, article 1. *www.newadvent.org/summa*

St AUGUSTINE: *The Literal Meaning of Genesis*, trans. John H. Taylor. Newman Press, 1982.

AVERROES: *The Philosophy and Theology of Averroes. Tractacta*. Manibhai Mathurbhal Gupta, 1921.

Stanley M. AWRAMIK: 'Respect for stromatolites'. *Nature* 441, June 2006.

Claude BABIN: *Autour du catastrophisme*. Vuibert, 2005.

F. BECK and W. GODIN: *Russian Purge*. Hurst & Blackett, 1951.

Daniel BECQUEMONT: 'L'Église anglicane face à l'évolutionisme', in *Actes du colloque sciences humaines et religion*, EHESS, Paris, September 2007.

Daniel BECQUEMONT: Private communication, 2009.

Francesco BERETTA: 'Doctrine des philosophes, doctrine des théologiens et inquisition romaine au 17e siècle: aristotélisme, héliocentrisme, atomisme'. No. 42. *Bulletin de la MHFA, Vera Doctrina: zur Begriffsgeschichte der Lehre von Augustinus bis Descartes.* Germany, 2006.

J. L. BERGGREN and Nathan SIDOLI: 'Aristarchus' *On the Sizes and Distances of the Sun and the Moon*: Greek and Arabic Texts', *Archive for the History of the Exact Sciences* 61, no. 3, 2007, pp. 213–54.

Jean-Pierre BIBRING: *Mars, planète bleue?* O. Jacob, 2009.

Hans BIERI: *Der Streit um das kopernikanische Weltsystem im 17. Jahrhundert. Freiburger Studien zur Frühen Neuzeit*, 9. Peter Lang, 2008.

Roger BILLARD: 'L'astronomie indienne, investigation des textes sanskrits et des données numériques'. *Publications de l'EFED*, LXXXIII, 1971. École Française d'Extrême Orient.

Jacques BLAMONT: *Le Chiffre et le Songe, Histoire politique de la découverte.* O. Jacob, 2005.

Paul BOGHOSSIAN: *Fear of Knowledge: Against Relativism and Constructivism.* Oxford University Press, 2006.

Jean-Yves BORIAUD: *Galilée, l'Église contre la science.* Perrin, 2010.

Jacques BOUVERESSE: *Prodiges et vertiges de l'analogie. De l'abus des belles-lettres dans la pensée.* Raisons d'Agir, 1999.

Jacques BOUVERESSE: *Peut-on ne pas croire?* Agone, 2007.

Jacques BOUVERESSE: Private communication, 2010.

Peter BROWN: *The Making of Late Antiquity.* Harvard University Press, 1978.

Giordano BRUNO: *Opere Complete, Oeuvres complètes.* Les Belles Lettres, 2006.

Massimo BUCCIANTINI: *Galilée et Kepler, Philosophie, cosmologie et théologie à l'époque de la contre-réforme.* Les Belles Lettres, 2008.

Georges-Louis Leclerc BUFFON: *Histoire naturelle générale et particulère.* Imprimerie royale, 1749a. Vol. V.

Georges-Louis Leclerc BUFFON: *Histoire naturelle générale et particulère.* Impr. royale, 1749b. *Recherches sur le refroidissement de la Terre & des Planètes*, Vol. I, available on www.buffon.cnrs.fr/correspondance._

Georges-Louis Leclerc BUFFON: *Lettre LXXVII au Président de Brosses.* www.buffon.cnrs.fr/correspondance, 1760.

Georges-Louis Leclerc BUFFON: *Les époques de la Nature.* Éd. du

Muséum, 1962 [1779]. (New edition by Jacques Roger, including the manuscript).

Cardinal Caietano BISLETI: *Sacra Rituum congregatione eminentissimo et reverendissimo domino cardinali Caietano Bisleti relatore Urbis et orbis concessionis tituli doctoris et extensionis ejusdem tituli ad universam ecclesiam nec non officii et missae sub ritu dupl. de comm. doctorum pontificum in honorem S. Roberti S.R.E. card. Bellarmino e societate Iesu.* Institut Catholique de Paris, 1931. Bibliothèque de Fels, cote 109 073.

Thomas Crowder CHAMBERLIN: 'On the interior structure, surface temperature, and age of the earth'. *Science* IX, 1899.

Auguste COMTE: *Cours de philosophie positive.* J. B. Baillière et fils, 1864. Vol. II, 19e leçon, p. 6.

Nicolas COPERNICUS: *Introduction à l'astronomie de Copernic.* A. Blanchard, 1975. Introduced, translated and edited by H. Hugonnard-Roche, E. Rosen, J.-P. Verdet.

Nicolas COPERNICUS: *De revolutionibus orbium coelestium.* Norimbergae, J. Petreium, 1543a. English translation: *Six Books on the Revolutions of the Heavenly Spheres,* 2008. www.webexhibits.org/calendars/year-text-Copernicus.html

Nicolas COPERNICUS: *Six books on the revolutions of the heavenly spheres.* Johannes Petreius, 1543b. English translation and notes by Edward Rosen, Polish Scientific Publishers, 1978.

Baron Georges CUVIER: *Discours sur les révolutions de la surface du globe, et sur les changements qu'elles ont produits dans le règne animal.* Edmond d'Ocagne, 1830, p. 226. English translation: *A Discourse on the Revolutions of the Surface of the Globe, the changes thereby produced in the animal kingdom.* London: Whittaker, Treacher & Arnot, 1829.

Jeff CUZZI: 'The first movement'. *Nature* 448, 2007.

G. Brent DALRYMPLE: *The Age of the Earth.* Stanford University Press, 1991.

Charles DARWIN: *On the Origin of Species by Means of Natural Selection, or the Preservation of Favoured Races in the Struggle for Life.* John Murray, 1859.

Richard DAWKINS: *The Blind Watchmaker.* W. W. Norton & Company, 1986.

Isaac DE LA PEYRÈRE: *Lettre de La Peyrère à Philotime, dans laquelle il expose les raisons qui l'ont obligé à abjurer la secte de Calvin qu'il professoit, et le livre des Préadamites qu'il avoit mis au jour. Traduit en françois du latin imprimé à Rome, par l'auteur mesme.* A. Courbe, 1658.

Jean-Baptiste Joseph DELAMBRE: *Histoire de l'astronomie ancienne.* Gabay, 2005.

Philippe DEPONDT: Private communication, 2009.

Philippe DEPONDT and Guillemette DE VÉRICOURT: *Kepler: l'orbe tourmenté d'un astronome.* Éd. du Rouergue, 2007.

Ahmed DJEBBAR: *L'âge d'or de la science arabe.* Pommier, 2005.

Philip ENGLAND: 'John Perry's neglected critique Kelvin's age for the Earth: A missed opportunity in geodynamics'. *GSA Today* 17, January 2007.

A. S. EVE: *Being the Life and Letters of the Rt. Hon. Lord Rutherford.* Cambridge University Press, 1939.

Annibale FANTONI: 'Galilée en procès, Galilée réhabilité?' Saint-Augustin, Francesco Beretta édition, 2005. In *Problèmes historiques posés par la 'clôture' de la question galiléenne.*

Mgr Albert FARGES: *Études philosophiques pour vulgariser les theories d'Aristote et de saint Thomas et montrer leur accord avec les sciences.* Bertche et Tralin, 1909.

Egidio FESTA: *L'erreur de Galilée.* Austral, 1995.

Egidio FESTA: *Galileo, La lotta per la scienza.* Laterza, 2007.

Paul FEYERABEND: *Against Method.* New Left Books, 1975.

Israel FINKELSTEIN and Neil Asher SILBERMAN: *The Bible Unearthed: Archaeology's New Vision of Ancient Israel and the Origin of Its Sacred Texts.* Free Press, 2001.

Israel FINKELSTEIN and Neil Asher SILBERMAN: *David and Salomon: In search of the Bible's sacred kings and the roots of western tradition.* Free Press, 2006.

Camille FLAMMARION: *Astronomie populaire.* C. Marpon et E. Flammarion, 1881.

Sigmund FREUD: *Standard Edition*, Vol. XXI. Hogarth Press, 1961.

Gad FREUDENTHAL: *Science in the Medieval Hebrew and Arabic Tradition.* Ashgate, 2005.

John G. C. M. FULLER: 'The age of the earth: from 4004 BC to AD 2002'. Geological Society, London, Special Publications, 190, 2001.

GALILEO GALILEI: *Sidereus Nuncius.* Tomaso Baglioni, Venice, 1610. English translation: *Sidereus Nuncius or The Sidereal Messenger.* University of Chicago Press, 1989.

GALILEO GALILEI: Letter to Benedetto Castelli, 1613, *http://inters.org/Galilei-Benedetto-Castelli*

GALILEO GALILEI: 'Considerations on the Copernican Opinion', 1615, *www.marxists.org/reference/subject/philosophy/works/it/galileo.htm*

GALILEO GALILEI: *Dialogo sopra i due massimi sistemi del mondo*. Bonaventiræ et Abrahami Elzevir, 1632. English translation: *Dialogue Concerning the Two Chief World Systems: Ptolemaic and Copernican*. Modern Library, 2001.

Saadia GAON: *The Book of Doctrines and Beliefs*. East and West Library, 1946.

Jacques GAPAILLARD: *Et pourtant, elle tourne! Le mouvement de la Terre*. Seuil, 1993.

André GILLET: *Une histoire des marées*. Belin, 1998.

Owen GINGERICH: *The Book Nobody Read: Chasing the Revolutions of Nicolaus Copernicus*. Walker Publishing Company, 2004.

Peter GODMAN: *The Saint as Censor: Robert Bellarmine between Inquisition and Index*. Brill, 2000.

Peter GODMAN: *Histoire secrète de l'Inquisition: De Paul III à Jean-Paul II*. Perrin, 2009.

David GOODSTEIN and Judith R. GOODSTEIN: *Feynman's Lost Lecture*. Norton, 2000.

Martin GORST: *Measuring Eternity: The Search for the Beginning of Time*. Broadway Books, 2001.

Sylvain GOUGUENHEIM: *Aristote au Mont Saint-Michel. Les racines grecques de l'Europe chrétienne*. Seuil, 2008.

Stephen Jay GOULD: *Time's Arrow, Time's Cycle: Myth and Metaphor in the Discovery of Geological Time*. Harvard University Press, 1988.

Edward GRANT: *Science and Religion, 400 B.C. to A.D. 1550: From Aristotle to Copernicus*. Johns Hopkins University Press, 2006.

C. M. GRAY: 'The Identification of Early Condensates from the Solar Nebula'. *Icarus* 20, July 1973.

Nidhal GUESSOUM: *Réconcilier l'Islam et la science moderne, L'esprit d'Averroès*. Presses de la Renaissance, 2009.

Richard Wesley HAMMING: 'The Unreasonable Effectiveness of Mathematics'. *The American Mathematical Monthly* 87 (2), February 1980.

Hermine HARTLEBEN: *Jean-François Champollion 1790–1832, sa vie et son oeuvre*. Pygmalion, 1997.

Sir Thomas HEATH: *Aristarchus of Samos, the ancient Copernicus; a history of Greek astronomy to Aristarchus, together with Aristarchus' Treatise on the sizes and distances of the sun and moon: a new Greek text with translation and notes.* Oxford University Press, 1913.

Paul-Henri Thiry HOLBACH: 'Fossiles' (article). *Encyclopédie de Diderot et d'Alembert*, 1758.

James HUTTON: *Theory of the Earth with proofs and illustrations.* William Creech, 1795. Facsimile of the Edinburgh edition, Wheldon & Wesley, 1972.

Guy ISRAËL: *Le système solaire.* Encyclopedia Universalis, 1983.

Patrick N. Wyse JACKSON: 'William Thomson's Determinations of the Age of the Earth', in *Kelvin: Life, Labours and Legacy*. Oxford University Press, 2008.

Patrick Wyse JACKSON: *The Chronologers' Quest: The Search for the Age of the Earth.* Cambridge University Press, 2006.

Catherine JAMI: Private communication, 2008.

JEAN-PAUL II: Discours de Jean-Paul II à l'Académie pontificale des Sciences. www.vatican.va/holy_father/john_paul_ii/speeches/1992/october/documents/h_jp-ii_spe_19921031_accademia-scienz fr.html, 1992._

Vincent JULIEN: *Sciences 'Agent double'.* Stock, 2002.

Jean KEPLER: *Mysterium Cosmographicum*, 1596. English translation: *Harmonies of the World*. Forgotten Books, 2008.

Jean KEPLER: *Astronomia Nova.* 1609. English translation: *New Astronomy.* Cambridge University Press, 1992.

Iosif B. KHRIPLOVICH: 'The Eventful Life of Fritz Houtermans'. *Physics Today*, July 1992.

Étienne KLEIN: *Discours sur l'origine de l'univers*. Flammarion, 2010a.

Étienne KLEIN: 'De la relativité du relativisme'. *Agenda de la pensée contemporaine* 16, 2010b.

Arthur KOESTLER: *The Sleepwalkers: A History of Man's Changing Vision of the Universe*. Penguin Books, 1959.

Alexandre KOYRÉ: *The Astronomical Revolution: Copernicus—Kepler—Borelli*. Dover, 1992.

Alexandre KOYRÉ: *Études d'histoire de la pensée scientifique*. Gallimard, 1973.

H. KRIVINE et al.: *Âge de la terre*. Available on www.cndp.fr/themadoc/histoiredelaterre/presentation.htm, 2007.

Jean-Louis KRIVINE: 'Wigner, Curry et Howard, La déraisonnable efficacité des mathématiques'. In *Exposé au colloque ARCo'04*, Sciences cognitives. Université de Compiègne, December 2004. www.pps.jussieu.fr/~krivine/articles/arco.pdf._

Thomas Samuel KUHN: *The Structure of Scientific Revolutions*. University of Chicago Press, 1962.

Jean LACOUTURE: *Champollion, une vie de lumières*. Grasset, 1989.

Diogène LAËRCE: *Vies, doctrines et sentences des philosophes illustres, Livre I, Les Sept Sages*. Garnier-Flammarion, 1965.

Pierre Simon LAPLACE: *A Philosophical Essay on Probabilities* [1825]. Dover, 1995.

J. LASKAR and GASTINEAU: 'Existence of collisional trajectories of Mercury, Mars and Venus with the Earth'. *Nature* 459, June 2009.

Bruno LATOUR: *Science in Action: How to Follow Scientists and Engineers through Society*. Harvard University Press, 1987.

Max LEJBOWICZ: 'Aristote au Mont Saint-Michel. Les racines grecques de l'Europe chrétienne'. *Journal of Medieval Studies*, 2008.

Georges LEMAÎTRE: 'La culture catholique et les sciences positives'. In *Actes du VIe congrès catholique des Malines*, Vol. V, 1936.

Fabienne LEMARCHAND: 'Controverse: les premières traces de vie'. *La Recherche* 354, June 2002.

Bruno LEVALLOIS: Private communication, 2010.

Bertrand LEVERGEOIS: 'De l'exemple à l'alibi'. *Europe* (no. 937), May 2007.

Claude LÉVI-STRAUSS: *The Savage Mind*. University of Chicago Press, 1966.

Tony LÉVY: *Figures de l'infini*. Seuil, 1987.

Tony LÉVY: 'L'existence de dieu selon Maïmonide (1138–1204)'. In *Controverses religieuses aux XIIe et XIIIe siècles et enjeux contemporains*. Ville de Perpignan et le Centre théologique Ramon Llull du diocèse de Perpignan, avec le concours de l'Institut Catholique de Toulouse, October 2010.

Geoffrey E. R. LLOYD: *Greek Science after Aristotle*. W. W. Norton & Co., 1973.

Martin LUTHER: *Table Talk*, trans. William Hazlitt. Christian Classics Ethereal Library, 2004. www.ccel.org/ccel/luther/tabletalk.html

Benoît DE MAILLET: *Telliamed ou entretiens d'un philosophe indien avec un missionaire français sur la diminution de la mer*. Pierre Gosse, 1755.

Benoît DE MAILLET: *Telliamed, or Conversations between an Indian Philosopher and a French Missionary on the Diminution of the Sea*.

University of Illinois Press, 1968.

Moses MAIMONIDES: *A Guide for the Perplexed*. E. P. Dutton, 1904. http://oll.libertyfund.org/index.php?option=com_staticxt&staticfile=show.php%3Ftitle=1256&layout=html

Charles MALAMOUD: Private communication, 2009.

Jean-Michel MALDAMÉ: 'La théologie naturelle. Une alliance de la foi et de la raison'. *Connaître* 22–23, December 2005.

Gérard MANHÈS et al.: 'The major differentiation of the Earth at 4.45 Ga.' *Earth and Planetary Science Letters* 267, March 2008.

Farouk MARDAM-BEY: Private communication, 2009.

Martino MARTINI: *Histoire de la Chine*, Vol. I. C. Babin & A. Seneuze, 1692. Translated from Latin by Father Le Pelletier

Karl MARX: 'Difference between the Democritean and Epicurean Philosophy of Nature', in *Marx Engels Collected Works*, vol. 1. Lawrence & Wishart, 1975.

Karl MARX: 'Contribution to the Critique of Hegel's Philosophy of Law: Introduction', in *Marx-Engels Collected Works*, vol. 3. Lawrence & Wishart, 1975.

R. MAZUMDER and M. ARIMA: 'Tidal rhythmites and their implications'. *Earth-Science Reviews* 69, 2005.

Frank MICHAELI: *Textes de la Bible et de l'ancien Orient*. Delachaux & Niestlé, 1961.

Georges MINOIS: *L'Église et la science*. Fayard, 1991.

Georges MINOIS: *Histoire de l'athéisme*. Fayard, 1998.

Peter MOLNAR et al.: 'Kelvin, Perry et l'âge de la Terre'. *Pour la Science* 364, February 2008.

Seyyed Hossein NASR: *An Introduction to Islamic Cosmological Doctrines*. Belknap Press of Harvard University Press, 1964.

Seyyed Hossein NASR: *Science and Civilization in Islam*. Harvard University Press, 1968.

Zuraya Monroy NASR: *Giordano Bruno: 1600–2000*, chapter 'La muerte de Bruno, la condena de Galileo y la prudencia de Descartes'. Facultad de filosifía y letras, universidad nacional autóma de México, 2002. Eds. Laura Benitez and Jose Antonio Robles.

Otto NEUGEBAUER: *The Exact Sciences in Antiquity*. Princeton University Press, 1952.

Otto NEUGEBAUER: *A History of Ancient Mathematical Astronomy*. Springer, 1975.

Isaac NEWTON: *The Correspondence of Isaac Newton*, volume II.

Cambridge University Press, 1960. (January 1680. From the original in King's College Library, Cambridge. In reply to Letter 246.)

Isaac NEWTON: *Philosophiæ Naturalis Principia Mathematica*. 1687.

Isaac NEWTON, 'Seven Statements on Religion', *www.newtonproject.sussex.ac.uk/view/texts/normalized/THEM00006*

Isaac NEWTON, 'Untitled treatise'. Yahuda MS 1. *www.newtonproject.sussex.ac.uk/view/texts/normalized/THEM00135*

Robert Russell NEWTON: *The Crime of Claudius Ptolemy*. Johns Hopkins University Press, 1977.

Sylvie NONY: Private communication, 2010a.

Sylvie NONY: *Abu l-Barakat al-Bagdadi: une théorie physique de la variation du mouvement au XIIe siècle, à Bagdad*. Doctoral thesis, Université Paris-Diderot Paris 7, 2010b.

Jonathan O'NEIL et al.: 'Neodynium 142 Evidence for Haden Mafic Crust'. *Science* 321, 2008.

Nicole ORESME: *Le Livre du ciel et du monde*. University of Wisconsin Press, 1968.

Cardinal Eugenio PACELLI: 'Lettres apostoliques Providentissimus Deus'. Proclamation of St Robert Bellarmine as Doctor of the Church, September 1931.

William PALEY: *Natural Theology; or, Evidences of the Existence and Attributes of the Deity*. 1802.

Anton PANNEKOEK: *A History of Astronomy*. Dover Books, 1989.

John PERRY: 'On the Age of the Earth'. *Nature* 11, 1895.

Isaac DE LA PEYRÈRE: *I preadamiti*. Quodlibet, Macerata, 2004.

Pascal PICQ: *Le Monde a-t-il été créé en sept jours?* Perrin, 2009.

Virgile PINOT: *La Chine et la formation de l'esprit philosophique en France (1640–1740)*. Paul Geuthner, 1932. Downloadable on classiques.uqac.ca/classiques

Henri POINCARÉ: *The Value of Science*. Modern Library, 2001.

Henri POINCARÉ: *Calcul des probabilités*. Jean Gabay, 1987.

Cardinal POUPARD: *L'affaire Galilée*. Éd. de Paris, 2005.

Claudius PTOLEMY. *Ptolemy's Almagest*. Princeton University Press, 1998.

Tariq RAMADAN: *Islam, the West, and the Challenge of Modernity*. Islamic Foundation, 2001.

K. RAMASUBRAMANIAN, M. D. SRINIVAS and M. S. SRIRAM: 'Modification of the earlier Indian planetary theory by the Kerala astronomers (c. 1500 AD) and the implied heliocentric picture of planetary motion'. *Current Science* 66, 1994.

Roshdi RASHED and Régis MORELON: *Histoire des sciences arabes.* Seuil, 1997.

Jeff RASKIN: 'A reply to Eugene Wigner's paper, "The unreasonable effectiveness of mathematics in the Natural Sciences" and Hamming's essay, "The unreasonable effectiveness of mathematics"'. *Akadeemia* 7, 2007.

Joseph RATZINGER and Paolo Flores D'ARCAIS: *Est-ce que Dieu existe? Dialogue sur la vérité, la foi et l'athéisme*. Payot, 2005.

Pietro REDONDI: *Galilée hérétique.* Gallimard, 1985.

Jean-François REVEL: *Histoire de la philosophie occidentale, de Thalès à Kant*. Nil, 1994.

Aimé RICHARDT: *Saint Robert Bellarmin 1542–1621, Le défenseur de la foi*. François-Xavier de Guibert, 2004.

Pascal RICHET: *A Natural History of Time*. University of Chicago Press, 2009.

Ana RIOJA and Javier ORDÓÑEZ: *Teorías del universo. De los Pitagóricos a Galileo*. Síntesis, 2004.

Robert H. ROMER: 'Foucault, Reich, and the mines of Freiberg'. *American Journal of Physics* 51, 1983.

Edward ROSEN: 'Calvin n'a pas lu Copernic'. *Revue de l'histoire des religions* 182, 1972.

Edward ROSEN: *Three Copernican Treatises*. Dover Books, 2004.

Micah ROSS: 'La circulation des savoirs astronomiques dans l'Antiquité'. *Pour la Science* 373, 2008.

François ROTHEN: *Et pourtant, elle tourne!* Presses polytechniques et universitaires romandes, 2004.

Carlo ROVELLI: *Anaximandre de Milet ou la naissance de la pensée scientifique.* Dunod, 2007.

Bertrand RUSSELL: *Science and Religion*. Home University Library, 1935.

Shlomo SAND: *The Invention of the Jewish People*. Verso, 2009.

George SARTON: *Ancient Science Through the Golden Age of Greece.* Dover Publications, 1993.

Youssef SEDDIK: Private communication, 2009.

David SÉNÉCHAL: 'Histoire des Sciences'. In *Notes de Cours. Université de Sherbrooke*, 2004. www.librecours.org/documents/20/2079.pdf.

William SHEA: *Galileo's Intellectual Revolution: Middle Period 1610–1632.*

Science History Publications, 1977.
Oscar SHEYNIN: 'The Treatment of Observations in Early Astronomy'. *Archive for History of Exact Sciences* 46:2, 1993.
Brian C. SHIPLEY: 'John Perry and the Age of the Earth', in *The Age of the Earth: From 4004 BC to AD 2002*, eds C. L. E. Lewis and S. J. Knell. Geological Society, London, Special Publications, 2001.
Arkan SIMAAN: 'Giordano Bruno, quatre siècles après'. *Science et pseudo-science* 28, October 2009.
Gérard SIMON: *Sciences et histoire*. Gallimard, 2008.
Colette SIRAT: *La philosophie juive au Moyen Âge: selon les textes manuscrits et imprimés*. Éditions du Centre national de la recherche scientifique CNRS, 1983.
Alan SOKAL: *Beyond the Hoax: Science, Philosophy and Culture*. Oxford University Press, 2008.
Alan SOKAL and Jean BRICMONT: *Intellectual Impostures*. Profile Books, 1998.
Steven SOTER: 'Les systèmes planétaires sont-ils pleins à craquer?' *Pour la Science* 365, March 2008.
Etienne SOUCIET: *Observations mathématiques, astronomiques, géographiques, chronologiques et physiques, tirées des anciens livres chinois ou faites nouvellement aux Indes, à la Chine et ailleurs, par les Pères de la compagnie de Jésus*. Rollin, 1732.
Pierre SOUFFRIN: 'Oresme, Buridan, et le mouvement de rotation diurne de la terre ou des cieux'. In *Terres médiévales*. Klincksieck, 1993.
Pierre SOUFFRIN: 'La théorie des marées de Galilée n'est pas une théorie fausse'. *Épistémologiques* 1–2, 2000.
Benedict de SPINOZA: 'A Theologico-Political Treatise', in *The Chief Works of Benedict de Spinoza*. Dover, 1951.
Richard STAUFFER: 'Calvin et Copernic'. *Revue de l'histoire des religions* 179, 1971.
Richard STAUFFER: *Avant, avec, après Copernic*. In *L'attitude des réformateurs à l'égard de Copernic*. A. Blanchard, 1975.
Jean-Jacques SZCZECINIARZ: *Copernic et la révolution copernicienne*. Flammarion, 1998.
Sir William THOMSON: 'On the Age of the Sun's Heat'. *Macmillan's Magazine*, vol. 5, March 1862.
Sir William THOMSON: 'On the Rigidity of the Earth'. *Royal Society of London Philosophical Transactions* Series I, 153, 1863a.

Sir William THOMSON: 'On the Secular Cooling of the Earth'. *Philosophical Magazine* 25, January 1863b.

James USSHER: *The Annals of the world, deduced from the origin of time and continued to the beginning of the emperour Vespasians reign, and the totall destruction and abolition of the Temple and commonwealth of the Jews, containing the historie of the Old and New Testament, with that of the Macchabees, also all the most memorable affairs of Asia and Egypt and the rise of the empire of the Roman Caesars under C. Julius and Octavianus …* J. Crook and G. Bedel, 1658.

Pierre VALIRON and Philippe DETERRE: *Chercheurs en science, chercheurs de sens*. Éd. de l'Atelier, 2009.

J. P. VANYO and S. M. AWRAMIK: 'Stromatolites and Earth-Sun-Moon Dynamics'. *Precambrian Research* 29, 1985.

Jean-Pierre VERDET: *Une histoire de l'astronomie*. Seuil, 1990.

Ilan VARDI. *Archimedes, the Sand Reckoner*. www.lix.polytechnique.fr/Labo/Ilan.Vardi/sand_reckoner.ps. 2007.

Jean-Pierre VERNANT: *L'univers, les dieux, les hommes*. Seuil, 1999.

Paul VEYNE: *Writing History: An Essay on Epistemology*. Manchester University Press, 1984.

Paul VEYNE: *Did the Greeks Believe in Their Myths?* University of Chicago Press, 1988.

Tony VOLPE: *Science et théologie dans les débats savants de la seconde moitié du XVIIe siècle*. Brepol, 2008.

Alfred WEGENER: *Die Entstehung der Kontinente und Ozeane*. Vieweg, 1915.

Steven WEINBERG: *The First Three Minutes: A Modern View of the Origin of the Universe*. Basic Books, 1977.

Eugène WIGNER: 'The Unreasonable Effectiveness of Mathematics in the Natural Sciences'. *Communications in Pure and Applied Mathematics* 13 (1), February 1960.

George E. WILLIAMS et al.: 'No heliotropism in Neoproterozoic columnar stromatolite growth, Amadeus Basin, central Australia: Geophysical implications'. *Palaeogeography, Palaeoclimatology, Palaeoecology* 249, 2007.

Nicolas WITKOWSKI: 'J.-B. Biot: un homme, une météorite'. *La Recherche* 326, December 1999.

Index

Afghani, Jamaluddin, viii–ix
alchemy, 51, 94, 95, 222n28, 239–240n70
Al-Khawârizm, 75
Almagest. See Syntaxis mathematica (Ptolemy)
Alvarez, Luis Walter, 26
Apollonius of Perga: and decentring of Earth, 223n37; and heliocentric explanation of model, 69–71; and model of as foundation for Ptolemy, 68
Appendix A, the Proofs of Earth's Motion, 171–178
Appendix B, Kelvin's Model and Calculation, 179–181
Appendix C, Radioactivity, 183–189
Appendix D, The Copernicus/Tycho Brahe Equivalence, 191–193
Appendix E, The Relativity of Trajectories, 195–196
Arabic astronomy, 75–76
Aristarchus of Samos: and Earth-Moon distance, 107–108, 108 fig. 33, 109 fig. 34; and Earth-Sun distance, 108–109; heliocentric model of, 67, 107, 114; possible persecution of, 67
Aristotle, 94, 103, 204n5, 204n7, 221n16; and atomism, 240n72; banning of teaching of, 237n54; break from thought of, 79, 80, 86, 93, 96, 127, 150, 230n1; and circular trajectories of planets, 114; and Earth's round shape, 63; fossilization of thought of, 223n30; and friction, 96–97, 140–141; and gravity, 80, 153; and homocentric spheres, 64, 67, 78; and immutability of Earth, 6; influence of, 5–6, 49, 64–65; and interest in Earth, 226n26; and Moon's shape, 119; and notion of two physics, 83, 86, 93, 96; physics of, 152; and a rational church, 128; and speed as proportionate to force creating it, 51; and supralunary and sublunary worlds, 6, 61, 64, 64–65, 78, 96–97, 204n8
Âryabhata, 62, 76
The Assayer (Galileo), 114, x–xi
astrology, 56; applications of, 51–52; in China, 61; and interest in planetary retrogradations, 55; prohibitions of, 52
astronomy, 14; and abandonment of geocentrism and development of heliocentric models, 48–49; of ancient India, 62; in Assyrian period, 61; and astrology, 51–52, 55–56; and Astronomical Unit, 81, 110; and Babylonian contributions, 59–61, 75; and beginning of quantitative astronomy, 71; and British astronomer James Bradley, 137, 138; and Cassini's estimates of Earth's distance from Sun, 109; and circulation of knowledge between cultures in antiquity, 59; and Copernicus on problems with Ptolemy's model, 85; and Danish astronomer Ole Christensen Rømer, 246n90; and dating of historical events, 154; and developments in ancient Greece, 62–75, 105; and diurnal motion of Earth and heliocentrism, 114; and Earth-Moon distance measured by Lalande and La Caille, 106; and Earth's age, 44–46, 154; and Earth's place and motion, 98, 154; and Egyptian scientists, 59–60; and formation of a star and planet system, 41–42; and formation of Earth's Moon, 30; and Galileo, 7, 105; and George Darwin's study of Earth-Moon dynamic, 22; and Ibn al-Shâtir's multi-epicycle model, 76; and instruments for observing the skies, 53, 71–72, 76–77, 78, 94, 105, 106; and invention of telescope, 94, 95, 119; and Irishman Samuel Molyneux, 138; and Kepler's number of planets and distances from Sun, 87–90; and last geocentric model by Tycho Brahe, 77–78; and major political role in China, 61; and measurements of the

ancients, 24, 48; and model of Claudius Ptolemy, 73–74, xvi; and model of Nicolaus Copernicus, xvi; and parallax (or triangulation) method for distances, 105–106, 106 fig. 31, 108, 109; and phases of Venus, 148; and Plato's model of the universe, 63–64; in Renaissance, 104; and retrogradation of planets, 48–49, 55, 56–59, 64, 79, 82; and solar system, 48, 56; and Thales of Miletus, 63; and Titius-Bode law, 110–111, 230n5; and tradition of Aristotle, 64–65; and use of instrument *dioptra*, 71–72. *See also* Galileo; Halley's Comet; heliocentrism; Hubble, Edwin; Kepler, Johannes; meteorites; Newton, Sir Isaac; planets; Ptolemy, Claudius; solar system
Augustine, St, 122, 209n5, 232n30
Averroes, 124; and creation, 6–7; and reading the Koran, 123, 130, 240n67
Awramik, S. M., 23
Babylonian astronomy: and astronomical observations of the Sun, Earth and Moon, 59–60; as contemporary of Egyptian civilization, 60; and data on clay tablets, 6, 60–61; and investigations into Mercury and Venus, 69; and mathematical models of solar system, 60–61; and names of constellations, 53; and predictions of eclipses, 60, 222n20; and Ptolemy's manipulation of sources, 75
Bacon, Francis, 7, 205n15
Barberini, Maffeo. *See* Pope Urban VIII
Becquerel, Henri, 33
Bessel, Friedrich Wilhelm, 139
Bethe, Hans, 39
The Bible, 29, xiv; and allegorical nature of Scripture, 122–123, 125, 127, 232n30; and book of Joshua, 114–115, 239n66; break from literal reading of, 79, 113; and canonical date of the Flood, 207n26; and Catholic view of Genesis, 128; and chronology of early genealogies, 8; and contradictions with historical facts, 1, 8, 10, 44; and developments in printing, 7; as divine word, 44, 113, 237–238n55, x; and Earth at centre of universe, 117; and Earth's age, 1, 18, 151; and Galileo on authority of Scripture, 120–122; and genesis of Earth, 6, 8, 14, 44; and Hebrew Bible, 10, 206n21; and immobility of Earth, 114, 230n2; interpretation of, 114, 125, 128–132, 131; and Israel, 10; and Jewish law, 239n63; and Mosaic chronology, 9–10; and need for a priest to understand Latin of, 125–126; and relationship of scientists to, 126–129; and story of the Flood, 1, 6, 8–11, 13–15, 24, 43, 215n45, xv; and Talmud, 130, 131; and Vulgate or Septuagint versions, 8, 9–10, 11, 205–206n21. *See also* Catholic Church; Christian world; Jewish world
The Bible Unearthed (Finkelstein and Silberman), 10, 207–207n34
Big Bang theory, 2, 3, 40, 45
Boghossian, Paul, 161,162, 166
Brahe, Tycho, 76, 84, 91 fig. 29; and belief in astrology, 51; contributions of, 78, 115; and end of 'corporeal spheres', 78; and Kepler, 77, 78, 87; and last geocentric model, 77–78; model of, 62, 77 fig. 25, 78, 117, 191; and Tycho's Nova, 78; and Uraniborg observatory, 78, 90
Bruno, Giordano, 16, 29, 133, ix; and conception of universe, 116; death and torture of, 115–116, 132; programme of, 93
Buffon, Georges-Louis Leclerc, 14–15, 16, xv; on biblical Flood, 209n10, 210n11; calorimetric experiments of, 18; and description of his experiments, 212n20; and Earth's age, 152; as father of scientific dating, 18; and model of sedimentation, 18; and times of sedimentation, 17
Buridan, Jean, 6
Burnett, Thomas, 13

catastrophism, 24, 27, 215n45
Catholic Church: and ban on Copernican doctrine, 118, 132; and battle with Galileo, 113, 118, 132–136, 155, 222n22, x, xvi; and canonization of Cardinal Bellarmine, 132–133, 134, 230n3, 241n75; and Cardinal Barberini, 119, 134–135; and condemnation of Galileo, 134, 241–242n83; and condemnation of heliocentrism, 134; and Congregation of the Index, 125; and Counter-Reformation, 114, 122; danger of questioning authority of, 135; and hidden God of Georges Lemaître, 128; and Index of prohibited books, 137; and influence of Averroes (Ibn Rushd), 123; and Inquisition, 115, 125, 238n57, xvi; and interview between Galileo and Barberini, 134–135; and missions in China, 10; and official text of Bible, 206n21; and Pope Alexander VII, 8; and Pope Benedict XIV, 132, 137; and Pope

Benedict XVI, 126, 128, 134, 217n59, 235–236n52; and Pope John-Paul II, 129, 133, 136; and Pope Urban VIII, 132, 133, 136, 155; and primacy of sacred texts, 134–136; and prohibition on Galileo's teaching, 136; scope of, 238n59; and trial and death of Giordano Bruno, 115–116; as well-established hierarchy, 129
Chadwick, James, 39
Chamberlin, Thomas Crowder, 20–21, 45
Champollion, Jean Francois, 10–11
chemistry: and Earth's age, 1, 33–34; and separation from alchemy, 51, 95, 222n28
China, x; astronomy of, 61–62; Catholic missions in, 10; and chronology of Chinese dynasties, 9–10; and discovery of precession of equinoxes, 61; and Jesuits' introduction of astronomical knowledge to court, 61; and the *Spring and Autumn Annals*, 9
Christian world: and Bible as source of knowledge, 1; and Counter-Reformation, 6, 114, 122; and Enlightenment, xiii, xiv; and intelligent design, 127–128, 153, 156, 234–234n50, xiv; and Jesuits, 9–10, 61; in Middle Ages, 5–6, 67, 104; and 'prime mover', 226n24. *See also* Catholic Church; creationism; religion; *The Bible*
Commentariolus (Copernicus), 79–80
convection, 35, 36
Copernican Revolution, 48–49
Copernicus, Nicolaus, 74, 95, 231–232n12; and Aristotle, 249n16; and belief in astrology, 51; and centrifugal force leading to stars' dispersion, 83; on composition of the planets as that of Earth, 83; criticism of, 103, 133; and distances of planets, 89–90, 225n3; and geometry, 117; and heliocentric theory in *De revolutionibus orbium coelestium*, 79, 81, 86; and heliocentrism, 7, 49, 79, 119, 145, 146; and his summary in *Commentariolus*, 79–80; and knowledge of models from Arab world, 76; mathematics of, 99; model of, 68, 77, 79–86, 98, 115, 117, 136, 141, 191, xvi; motivations of, 85, 147; and planets' movements, 97; and planets' speeds in relation to distance from Sun, 90; postulates of, 80–82; reading texts of, 103; and rejection of Earth's centrality and immobility, 85–86; summary of contributions of, 86; and system of epicycles, 79, 84, 91
Coriolis force, 119, 143, 144, 246n87; and Coriolis effect's role in tides, 232n20; discovery of, 154
Counter-Reformation, 6, 114, 122
creationism, xiv; and *The Bible* as source of knowledge, 44; and Creation, 3, 6, 8, 13–14, 17, 209n5, 236n52; and date of second coming of Christ, 207n31; and George Bush, 29, 217n60; and intelligent design, 235–236n50; and Renaissance, 156; retrograde movement accompanying, 157; revival of, vii
The Crime of Claudius Ptolemy (Newton), 75
Curie, Marie, 33, 250n28
Curie, Pierre, 33, 218n9, 226n23
Cuvier, Georges, 17; and belief in biblical Flood, 15; as founder of catastrophism, 24, 215n45; and leaving terrain of science, 156

Darwin, Charles Robert, xv; and calculation of age of hills of the Weald in Kent, 17, 27, 28; and end of belief in intelligent design, 156; and gradualism, 215–216n53; and hypothesis of Sun transferring magnetic energy which heated Earth, 28; influence of Charles Lyell on, 26; and linkage of origins of humans with that of animals, 86; and missing links, 27–28; and natural selection, 153; and rejection of Kelvin's figures for Earth's age, 27–29, 104, 154; relevance of, 157; scandal over work of, 136; and support of by Ludwig Boltzmann, 217n55; and theory of evolution, 27–29, 43, 45, xvi
Darwin, George, 22
deferents, 70–72, 78, 79; decentring of by Ptolemy, 74; definition of, 69
Democritus, 5
De revolutionibus orbium coelestium (Copernicus), 79–82, 86, 151
Descartes, René, 8, 15, 209n2, 249n29, viii; and decision not to publish book on heliocentrism, 127; exile of, 136; and laws of physics, 13; and vortices to explain electromagnetic waves, 98, 140
dinosaurs: disappearance of, 2–3, 26; and discovery of the archaeopteryx, a link with birds, 27–28; and impact of meteorite causing their disappearance, 26; uncertainty about cold-bloodedness of, 216n51
diurnal motion: of planet Earth, 14, 61, 62,

68, 81, 136, 137, 140, 145–146; of planets, 55; of stars, 53, 145
Doppler effect, 40, 139, 174
Doppler-Fizeau effect, 139–140, 139 fig. 35, 144, 175
Doubts on Ptolemy (Ibn al-Haytham), 76

Earth: and absolute dating of rocks, 37–38, 219n20; accretion period of, 43; age of, 6–8, 15, 17, 25, 35–36, 40, 104, 131, 148, xiv, xv; age of as 4.55 billion years, 1, 18–21, 33, 43–46, 166,
Earth (*continued*)
xiii, xvi; and age of calculated by B. B. Boltwood, 38; and age of calculated by Charles Darwin, 27; and age of calculated by John Perry, 35, 36; and age of calculated by Sir William Thomson Kelvin, 28, 35, 36, 38; and age of deduced from meteorites, 41–43; and ancient structures called stromatolites, 2, 23, 213–214n42; and attempts at dating, 16, 34, 37–43, 211n13, xvi; authority over age of, 151; and axis of the poles, 23–24, 54 fig. 11, 56; and biblical Flood, 1, 6, 8–11, 13–15, 24, 43, xv; and carbon-14 dating, 220n28; and centrifugal force due to diurnal rotation, 140; and changing distance of the Moon, 15, 22; and Chicxulub crater in Gulf of Mexico, 26; and clocks for measuring age of, 1, 13, 15–24, 33, 37–38, xv; and collision with Theïa to produce the Moon, 30; and composition of as molten rock, 18; and convection, 35, 36; cooling of, 18, 29, 35–36; and dating of strata by fossils, 17, 213–214n43; death of, 1, 29–30, 31 fig. 7; diagrammatic temperature curves for, 36 fig. 10; dimensions of, 49; and distance to Moon, 106, 154; and distance to nearest stars, 106; and distance to Sun, 49, 73, 81, 105–107, 108–109; diurnal motion of, 55, 61, 62, 68, 81, 136, 137, 140, 142, 144, 145–146, 232n17; and Earth-Moon dynamic, 22, 49, 53–55; ellipse as explanation for motion of, 73; and equations for proof of motion around Sun-the aberration of the stars/light, 171–173; equator of, 144; and equilibrium between erosion and sedimentation, 25; and equinoxes, 7, 15, 23–24, 56, xvi; evolution of ideas about, 153; and extinctions of flora and fauna, 216n49; and flattened shape at poles, 94, 140, 145; and floods, 15, 24; formation of, 2, 3, 43; and Foucault's pendulum in 1850, 137; and Fourier's equation, 19; and global warming, 213n32; and gravity, 22, 94; and Greek mythology, 5, 65; and heat spreading by convection, 35; and heliocentric theory in *De revolutionibus orbium coelestium*, 80–86; as immobile and centre of universe, 65–66, 81, 85, 113, 114, 117, 135, 146, 153, x; and impossibility of an absolute age for, 154; initial conditions of, 18, 19; and interior of behaving like a fluid, 35; and length of days and years, 23, xvi; and limit conditions, 18–19; as a magnet, 93; mantle of, 35, 36; marine life on, 210–211n15, 214n43; and measurements of in nineteenth century, 98; and model of Apollonius of Perga, 68–71; and model of Hipparchus, 72; and model of sedimentation, 18; and motion of as a standard of time, 100; and movement around Sun, 55, 56 fig. 13, 114, 118, 142, 146; movement of, 30, 41, 44, 47–48, 58 fig. 16, 96, 106, 136, 148, 154, 239n66, xiv–xv, xvi; and new source of energy, 33–35; nuclear origin of internal heat of, 185–189; and orbit around Sun in an ecliptic plane, 23–24, 54 fig. 11, 56; orbit of, 30, 84, 91–92, 109, 142; origins of, 5–7, 29, vii, xv; and presence of iridium, 26, 27; and proof of motion around Sun-the aberration of the stars/light, 137–138; and proofs of its motion, 137–146; radius of, 106, 166, xiii; and reason for seasons, 55–56; and Reich's experiment of 1831, 137; and relationship between the length of day and tide cycle, 22–23; rigidity of, 35; and rocks as photographic plates, 39–40; and rotation on itself, 23, 55, 81–83, 86, 114, 118, 123, 140, 142, 144, 145–146; and salt content of oceans, 15, 21, 34; size of, 106; and solstices, 56, 63; spherical shape of, 63; stratification and erosion of, 15–16, 17, 23, 25, 27; and temperature gradient at surface, 19, 35, 36; and temperature of for development of life, 18–19, 252n3; terrestrial crust of, 35, 36, 41; and theory of chaos, 30; and thermic death, 29–30; and tides, 19, 22–23, 93, 118, 214n42, 232n17, 232n20; time taken for formation of, 42, 43; and trade winds caused by Coriolis effect, 119, 145; and trajectory of as an ellipse, 73, 81, 90–92, 100, 166, xiii; and upper

limit of its age, 20; and variation of the period of a pendulum along meridian, 137, 140; and volcanic eruption in India, 26–27. *See also* Aristotle; creationism; ellipses; equinoxes
eclipse, 7, 53, 60, 61, 63, 147; diagram of lunar eclipse, 54 fig. 11; used to deduce Moon's size, 107–108
Egyptian civilization, 10–11, 59–60, 63
Einstein, Albert, 150; famous equation of, 183; and Hiroshima, 103; mathematics of, 99; and theory of relativity, 98
elements: and activity of a radioactive element, 37; and argon, 39; and disintegration chains, 39–40, 42–43; and heavy elements such as lead or uranium, 41; and helium, 34, 38, 218n2, 218n12; and iridium, 26, 27; and isochrones method for verifying disintegration chains, 39, 40, 189; and lead, 39, 41, 42; and period of a radioactive element, 37, 38, 44; and potassium, 39; and presence of radioactive elements in the ground, 34; and radioactive elements as a source of heat, 34; and radium, 33–35; and rubidum, 38, 39; spontaneous splitting of, 40; and strontium, 39; and thorium, 39; and uranium, 33, 40, 41, 42; and uranium disintegration into lead, 37–38, 39
ellipses: and discovery of elliptical trajectories, 90–92, 94; and Earth's elliptical orbit, 142; Sun as cause of motions of, 92
epicycles, 69, 79, 90, 91, 117
equinoxes, 56 fig. 13, 57 fig. 14, xvi; diagram of, 56 fig. 13; precession of, 7, 23–24, 61, 72, 154, 205n18, 221n7, 222n27
Eudoxus, 65–67
Europe, 61; and diagram of scientists of Middle Ages, xxi; and knowledge from France, 10, 33, 136, xvi–xvii; and knowledge from Greeks, 59; and Middle Ages corresponding to a golden age in Islamic world, 75; and reason and faith during the Middle Ages, 103–104; and use of Bible in Christian Europe, 1. *See also* Renaissance
evolution: and Church of England, 29; and long timeframe of physics, 153; of planet Earth, 1; theory of, 27–29, 43, 131, 152, xvi. *See also* Darwin, Charles Robert
The Evolution of Species (Darwin), 27
fossils: biological nature of, 16, 17; and dating of Earth's strata, 17; and fossilization, 27; as proof of time passing, 25; and verification of the Flood, 13, 24
Foucault's pendulum, 137, 143–144, 145, 177–178
Fourier, Joseph, xv; equation of, 19, 20, 35, 179–180; theory of, 34
Fourier's equation, 19, 20, 35, 179–180
France: and Bequerel family at Paris Museum of Natural History, 33; and Galileo's doctrine, 136; and Louis XIV, 10; and science, xvi–xvii

Galilei, Galileo, 16, 18, 29, 62, 93, xiv; and argument of trade winds to show Earth's rotation, 119, 136; astronomy and physics of, 7, 84, 146, 222n24; and atomism, 241n72; on authority of Scripture, 120–122, 124; and battle with Catholic church, 113–114, 132–136, 155, 250n24; biography of by Festa, 120; condemnation of, 114, 115, 132, 135–136, 145, 241n74, 242–246n83, x, xvi; and "Considerations on the Copernican Opinion", 120–121; and contradicting biblical assertions, 131–132; and Copernicus' ideas, 86, 117, xvi; criticism of, 103; and culture of Middle Ages, 122; and discovery of Jupiter's moons, 117; and discovery of phases of Venus, 78, 95; and Earth's movement proved by tides, 118–119, 133–134, 136; and fall of bodies, 97; and free fall of an object, 142–143; and friction, 141; and heliocentrism, 49, 67, 136–137; and inertia, 94; and interview between Galileo and Barberini, 134–135; on language of mathematics, 98, 114, xiv; as leading mathematician, 113, 150, x–xi; and measurement of parallax, 138–139; modern thinking of, 119–120; and position on astrology, 45–46; reading texts of, 103; and rejection of Kepler's ellipses, 103, 104; relevance of, 157; on salvation, 121; and texts by God vs. texts by Church Fathers, 122, 132; and theory of tides, 232n20; and threats of torture, 134; trial of, 136, 230n3, 251n33; and use of reading-glass from Holland, 53
geocentrism, 47–48; abandonment of, 80; and models of with earth at centre, 77–78
geology: and ages of rocks for determining Earth's age, 38, 39, 40–41; and biblical

teaching of the Flood, xv; and Earth's age, 1, 24, 27, 43, 44, 45; and long timeframes, 24, 152; and time for strata to sediment, 27. *See also* Ptolemy, Claudius
geometry: and abandonment of Eudoxus' model of spheres, 66–67; of Aristarchus, 107; and contributions of the Greeks, 49, 62–63; and Copernicus' ideas, 117; and geometrical framework for astronomic studies, 48, 49, 66–67; and Kepler, 87, 90, 92
gravity, theory of, 3, 153, xvi; and flattening of Earth as responsible for a reduction in gravity, 140; and Giordano Bruno, 115; and gravitational collapse, 41; and gravitational energy, 20; and Newton, 94, 96, 109–110, 147; and Newton's successors, 98, 110; and tides, 119
Greek antiquity, viii; and ancient Greece as cradle of modern science, 59; and Archimedes, 67; and astrology in ancient Greece and Rome, 52; astronomical developments of, 62–75; and atomism, 5, 240n72; and Dendera zodiac, 10; diagram of scientists of, xx; and Diogenes Laertius, 63; and Epicurus, 5; and estimations of relative sizes and distance of planets, 105; and Eudoxus' model of homocentric spheres, 65–67; and geometrical framework for astronomic studies, 48, 49, 61; and heliocentric explanation of Apollonius' model, 69–71; and Heraclides of Pontus' theory of epicycles, 69; and Hipparchus' model, 72 fig. 22; and Hipparchus of Nicea's improvement, 71–72; importance of prediction in, 62; and model of Apollonius of Perga, 68–71, 68 fig. 18, 72; and model of Aristarchus of Samos as first heliocentric model, 67, 107; and myths of Earth's creation and position, 5, 63; and physics of Aristotle, 152; and Plato's model of the universe, 63–64; and Pythagorean school, 63; and uncertain origins for discoveries, 62. *See also* Aristarchus of Samos; Aristotle; Ptolemy, Claudius

Halley, Edmond, xv; comet of, 97, 213n34, ix; and Earth's age, 21; and oceans' water being charged with river salts, 21; and parallax (or triangulation) method, 109; and solar parallax, 228–229n3
Halley's Comet, 97, 213n34, ix
heliocentrism, 7, 56 fig. 13, 150, 230n1; and aberration of light, 137, 171–173; and Aristarchus of Samos, 67, 107, 114; attributed to Pythagoras by church, 222n22; and doctrine established in eighteenth century, 137; and Doppler-Fizeau effect, 137; and Earth and other planets, 55, 80–86; and heliocentric explanation of Apollonius' model, 69–71; and Indian model of Somasutvan, 62; and introduction of to Chinese court by Jesuits, 61; and Kepler, 93; Luther's condemnation of, 118; and model of Aristarchus of Samos as first heliocentric model, 67; and model of Claudius Ptolemy, 119; and model of Nicolaus Copernicus, 79–86, 82 fig. 27, 119, 147–148, 191; papal condemnation of, 130, 132; and physical Sun as centre of world, 91; problems with, 152; and René Descartes, 127; and retrogradation of planets, 57 fig. 15, 82; and Sun's apparent trajectory against sky, 57 fig. 14, 81; and Tycho Brahe, 62. *See also* Aristarchus of Samos; Copernicus, Nicolaus; Jesuits; Kepler, Johannes; Newton, Sir Isaac
Hipparchus of Nicea, 71–72, 73, 105, 108, 223n37
History of China (Martini), 9
Holmes, Arthur, 40
hominids, 3; and Lucy, 2, 28; and Toumaï, 2, 28
Houtermans, Fritz, 40
Hubble, Edwin, 40
humans: ancestors of, 28; and Cro-Magnons, 2; and Darwin's linkage of origins of with that of animals, 86; and first appearance on Earth, 3; and natural selection, 152; and Neanderthals, 2; and reasons for studying the motions of the heavens, 51; and theory of evolution, 29
Hutton, James, 25–26

India: and Arabic numerals, 75; astronomy of, 62; volcanic eruption in, 26–27
inertia principle, 140–143, 144, 150
intelligent design, 234–235n50, vii, xiv; and religion, 127–128, 153, 156; and Sir William Thomson Kelvin, 29
internet, the, 203n1, 205n14, xix
Islamic world, 126, 129; and Arabic astronomy, 75–76, 224n44; and Arabic translation of Ptolemy's *Almagest*,

73; and Arab renaissance, 75, viii; and astrology, 52; and astronomers Nasir al-Din al-Tusi and Ali Qushji, viii; and Averroes (Ibn Rushd), 123, 130; and Avicenna, 123, 130, 211n15, 226n24; and creation, 6–7; diagram of scientists in golden age of, xxi; difficult circumstances of, vii, xi; and education, vii; golden age of, 130, 202n8, 238n57, xxi; and Ibn al-Shâtir, 76; and Ibn Yunus' astronomical tables, 75; and Islam, ix–x; and Jabir ibn Aflah in Spain, 76; and Jamâl ad-Dîn al-Afghânî and Mohammed Abduh, 130–131; and Koran hard to understand for Muslims, 126; and modern Islamic reformism, 130–131; and Nasir al-Din al-Tusi 's model of the *Tusi-couple*, 76, 223–224n48; and obscurantism, vii–viii; and openness of scholars in Baghdad, viii; and reason and faith during the Middle Ages, 104; and recent eccentricities of thought, ix; and revival of religious fundamentalism, xiv; and role of Maimonides, 125; and scholar Abu'l-Barakat al-Bagdadi, 64; and standard modern Arabic, 234n44. *See also* Afghani, Jamaluddin; Averroes; Koran; Pakistan

Jesuits: and acceptance of Brahe's model, 117; and diurnal motion of Earth and heliocentrism, 61; and Martino Martini, 9; and promotion of Septuagint Bible, 10

Jewish world: and Hebrew text, 209n3, 239n64, 239n65; and Israel, 10; and Jewish medieval philosophy, 130; and rejection of astrology, 52; and view of Saadia Gaon on science vs. sacred text, 124; and weight of religious tradition, 238n57; and work of Maimonides, 124–125

Jews, 129; attempts at conversion of, 8; and Hebrew text, 10, 130. *See also* Maimonides; *The Bible*

Joly, John, 21, 34, 39

Kasravi, Ahmed, ix

Kelvin, Sir William Thomson, 127, 183, xv; and absolute dating, 45; and Buffon's theory, 18; and calculation of Sun's age, 20, 36; and Earth's age, 18, 19–20, 28, 34–36, 38, 43, 151; as embodiment of scientific procedure, 44; and equation for heat, 18; and hypothesis of Earth's rigidity as incorrect, 35; and initial conditions, 181; and intelligent design, 29, 217n56; model and calculation of, 179–181; and Moon's age, 22; and universality of laws of physics, 208n1; and use of Fourier's equation, 35

Kepler, Johannes, 1, 7, 95, 132, 192, xvi; and astronomy, 8, 49, 74, 148; and belief in astrology, 51; and break with Aristotle, 80, 86, 93; and causes of planets' movement and number, 87–90; and challenge to constraint of only circular motion for planets, 65; and defense of Copernican system, 129; as disciple of Tycho Brahe, 77, 87; and discovery of planets' elliptical trajectories, 79–80, 90–92, 94, 104; and distances of planets relative to those of Copernicus, 89–90, 89 table 1; and divine will, 87; and Earth's age, 8; and exactness as manifestation of divine law, 90; and geometry, 87, 90, 92; and gravity, 98; and *Harmonices Mundi*, 92; and laboratory notebook *Astronomia Nova*, 87; laws of (or three laws of motion), 80, 90, 92, 93, 94, 96–98, 105, 109, 191; and laws of areas, 91–92; as leading mathematician, 113, x; and magnetic power, 93, 97, 104; and motive spirit of Sun, 92–94; and *Mysterium Cosmographicum*, 87, 90; and planets' speeds in relation to distance from Sun, 90–91; reading texts of, 103; religion of, 129; successive polyhedrons of, 87–89, 88 fig. 28, 97, 111; and Sun as cause of planetary motion, 84, 90–91; and Sun's magnetic power, 97; and texts by God vs. texts by Church Fathers, 122; on the tides, 118–119; and trust in Tycho Brahe's data, 90

Keynes, John Maynard, 94–95, 226n27

Koestler, Arthur, 228n46; and conservatism of modern science, 104; and division between reason and faith during the Middle Ages, 103–104

Koran, ix, xiv; Arabic of, 126, 234n44; and Avicenna and Averroes, 130, 240n67; as divine word, 237–238n55; and Earth's genesis, 6; interpretation of, 130–131, 209n3, 240n67; and Sun stopping, 239n66

Krivine, Hubert, 166–167, 169, vii

Krüger, Johann Gottlob, 14

La Peyrère, Isaac de, 8, 206n25, 231n12

Laplace, Pierre Simon, 100–101, 102, 149, 201n1
Latour, Bruno, 161, 162–165, 169, 250n30
Lemaître, Georges, 128
Le Verrier, Urbain, 97–98
Lévi- Strauss, Claude, 51
Lucretius, 5
Luther, Martin, 129–130
Lyell, Charles, 26, 43

Maillet, Benoît de: and Creation, 17; and hypothesis of Earth's origins from the sea, 16, 17; and precautions taken to avoid church's wrath, 16–17
Maimonides, 130; and allegorical nature of Scripture, 125; as against eternity of Earth, 7; and *Guide for the Perplexed*, 124–125; and rejection of astrology, 52
Malthus, Rev. Thomas Robert, 27–29
Martini, Martino, 9
Marx, Karl, 47, 201–202n1, 201n1, 204n4
mathematics, 75, xi, xiv; and algebra, 75, viii; and Al- Khawârizm's Arabic numerals, 75; and Anderson's discovery of electrons with positive charge, 99; and Appendices, xix; and Babylonian contributions, 60–61; and discovery of black holes, 99; Galileo on language of, 98, 114; and geometry, 48, 49, 63, 87, 92, viii; and Hertz's discovery of radio waves predicted by Maxwell's equations, 99; and introduction of trigonometric functions by Hipparchus, 72, 105; in Islamic world, 75; and Kepler's three laws, 97; and mathematizing of reality, 48; and physics, 48, 80, 98; and science, 237–238n56; and Sir Isaac Newton, 94, 99–101; and system of coupled equations, 100; and theory of universal movement and gravitation, xvi; and trigonometry, 108; and truth, 149; unreasonable efficacy of, 101–102; value of, 99. *See also* Âryabhata; Greek antiquity; Kepler, Johannes; Ptolemaic model
Mesopotamia, 59, 60, viii
meteorites, 27; and age of Earth, 41–43; and determining age of by lead-lead method, 42; and determining age of by other disintegration chains, 42–43; and disappearance of dinosaurs, 26; and meteor that caused Meteor Crater in Arizona, 42; in north of Mexico, 252n5; and rocks of as most contemporary with formation of Sun, 42; and uranium, 42; as violation of supralunary world, 204n8
Moon: age of, 22; and attempts at dating Earth, 22, xvi; and changing distance from Earth, 15, 22; diameter of, 107 fig. 32, 108; dimensions of, 49, 63; and Earth-Moon distance, 49, 106–108, 154; and eclipses, 7, 53, 54 fig. 11, 60, 107–108; epicycles of, 224n49; and Eudoxus' model of homocentric spheres, 66; and full moon, 53–55, 54 fig. 11, 61; and George Darwin's study of Earth-Moon dynamic, 22; and gravity, 22; and illumination of by the Sun, 53, 54 fig. 11, 108; mountains of, 95; movement of, 71, 80, 93; and new moon, 53, 54 fig. 11; phases of, 119; radius of, 106–108; and relationship with the Sun and Earth, 54 fig. 11, 60; as result of collision between Earth and Theïa, 30, 213n37; and separation from Earth, 22; and speed at which it moves away from Earth, 213n36; and supralunary and sublunary worlds, 6; and tides, 93, 213n39; and use of nautilus shells for determining lunar cycles, 23, 154, 214n43
Mysterium Cosmographicum (Kepler), 87

Newton, Robert Russell, 75
Newton, Sir Isaac, 29, 43, 79, xiii, xv; and acceleration of a body, 96, 98, 222n24; and alchemy, 94, 95; and astrology, 52; and astronomy, 8, 84, 94–98, 148, 205n18; and *The Bible*, 44–45, 126–127, 128; and biblical dating, 13–14, 23; and birth of modern physics, 80; and chronology of kingdoms, 206n20; and deduction of Kepler's laws, 94, 96; and divine will, 153, 234n51; and Earth's age, 8, 226n26; and flattening of Earth as responsible for a reduction in gravity, 140; and inertia principle, 141; and invention of infinitesimal calculus, 94; and invention of telescope, 94; and law of universal attraction, 80, 150, xvi; and law of universal gravitation, 94, 96, 109, 119, 147; mathematics of, 99–101; mechanics of, 48–49, 65, 96, 97, 149; and optics and light, 94; personality of, 94–95; reading texts of, 103; and theology, 94, 95; and theory of universal movement and gravitation, 80, 96, 97, xvi; and tides, 119; and unreasonable efficacy

of mathematics, 102. *See also Principia Mathematica* (Newton)
nuclear energy, 33–35, 185

obscurantism, 103, 136, 156, vii, viii, xiv
oceans: and dating, 34; salt content of, 15, 21
On the Heavens (Aristotle), 6
On the Origin of Species by Means of Natural Selection (Darwin), 151
On the Revolutions of the Heavenly Spheres (Copernicus), 117–118
orbit: of Jupiter, 221n9; and orbital zones, 111; of planet Earth, 23–24, 30, 43, 56, 84, 91–92, 109, 142; of planets, 59 fig. 17, 92, 93, 105, 109–111, 191–193; and survival of certain orbits, 249n18. *See also* eclipse; Ptolemy, Claudius
Oresme, Nicole, 123

Pakistan, x, ix
parallax of stars, 81, 137–139, 144, 146, 173; and first measurement of by Friedrich Wilhelm Bessel, 139; Galileo's failed measurement of, 138–139
Pascal, 13, 81
Pasteur, Louis, 104, 153
Patterson, Clair, 40–41, 42
Perry, John, 35; and role of convection, 45; and theory of Earth's age, 36
philosophy: and Aristotle, 5–6; and Earth's age, 1; and Galileo, 114; and Giordano Bruno, 115; and Laplace, 100–101; and Pascal, 13; of René Descartes, 15. *See also* postmodernism; science
physics, 75, x, xix, xv–xvi; and advances in atomic and nuclear physics, 39; and biblical time, 13; and birth of modern physics, 80; and creativity of vacuum, 151; and discovery of isotopes by Soddy, 39; and Earth's age, 1, 44, 46; and Galileo, 7; laws of, 1, 13, 15, 26, 28; and life of Fritz Houtermans, 218–219n14; long timeframe of, 153; and Ludwig Boltzmann, 217n55; and mathematics, 48, 98; and model of Nicolaus Copernicus, 82; and models of Earth, 113; and nuclear physics, 3, 183, x; and physicist Eugène Wigner; and quantum mechanics, 152; and quest for grand unification, 225n16; and Sir Isaac Newton, 94; and Sir William Thomson Kelvin, 43; and two physics of Aristotle, 65, 83, 86, 93, 96. *See also* Aristotle; Salam, Professor Abdus
Pinot, Virgile, 10
planets, 30, 69 fig. 19; and astrology of ancient Greeks, 52, 61, 63–64; and astrology of Babylonians, 61; and correlation between distances from the Sun and revolution periods, 86; and Creation, 3; and declassification of Pluto, 226n19; and discovery of Neptune, 97, 100, 110; and discovery of Uranus, 97, 110; and distance from Sun, 81, 86, 88–90, 109–111, 110 table 2; and distances of relative to Earth-Sun distance, 49; and distance to nearest stars, 49; diurnal motion of, 55; eastward movement of, 55; and figure of eight curve called the *hippopede*, 66; and formation of as process of accretion and destructions, 42, 111; and heliocentric movement, 55, 56 fig. 13, 79–80, 96; influence of, 22, 55; and interactions with one another, 100; irregularity in movement of, 67, 82, 90; and Jupiter, 42, 55, 60, 70, 95, 117, 119, 136; Latin names for, 55; and Mars, 55–57, 58, 58 fig. 16, 60, 78, 90–92; measurement of the period of, 92; and Mercury, 55, 58, 60, 69, 90, 91 fig. 29, 97–98, 223n39; and model of Apollonius of Perga, 70–71, 70 fig. 20, 71 fig. 21; and model of Claudius Ptolemy, 73–75, 76, 84, 86, 141, xvi; and model of Nicolaus Copernicus, 141; and motion caused by universal gravitation, 96, 98; and motion of as a circle, 96; and motive spirit, 65, 94; and notion of six planets, 87–88; number of, 87–90, 225n16; orbital dimensions of, 105, 111; orbital radii values of, 109–110; and orbital size, 92; orbital times for, 59 fig. 17; and orbits of according to Kepler, 93; and parallax (or triangulation) method for distances, 109; and perihelion, 91; and prediction of positions, 109–110; and relative sizes and distances of estimated by Greeks, 105; retrograde movement of, 48–49, 55, 56–59, 64, 79, 82; and Saturn, 42, 55, 58, 60; and speed of in relation to distance from Sun, 56, 82, 90; and Sun as cause of planetary motion, 84, 90–91; system of, 41; trajectories of, 97, 100–101, 141, 142; and trajectories as ellipses, 79–80, 84, 90–92, 94, 100, 103, 104; and Uranus and Neptune as unknown to ancients, 55, 226n19; and variation in brightness of Mars or Venus, 222n27; and Venus, 55, 58, 60, 69, 78, 95, 109, 119, 136, 148, 229n3. *See also* Earth; ellipses

Plato, 63, 64, 79, 114
Poincaré, Henri, 33, 101, 102, 136, 145, 149, 227n41
Pope Alexander VII, 8
Pope John-Paul II, 29
Pope Leo XII, 10
Pope Urban VIII, 132, 133, 136, 155; and condemnation of Galileo, 114, x; and keeping of a personal astrologer, 52
postmodernism, 160–161, 162, 166, 168–169, 250n30
Principia Mathematica (Newton), 96, 151
Ptolemaic model, 67, 74 fig. 24, 141, 148, xvi; and conformation to Bible, 114; critiques of, 76, 77–78, 84, 98, 146; discarding of, viii; and geocentric mathematical model, 73–74; predictive power of, 79; rehabilitation of, 145
Ptolemy, Claudius, 60, 68, 71, 72, xvi; astronomical observations of, 74–75; and boat moving against shore, 221n16; Copernicus' mention of, 83; criticism of, 76, 103, 115; and distances of planets relative to those of Copernicus, 225n3; as great astronomer and geographer of Hellenistic period, 73; mathematics of, 99; model of, 77–78, 84, 98, 114; and planets' movements, 97

radioactivity, 20, xv–xvi; and absolute dating of rocks, 37–38; and advertisements, 33, 34 fig. 9; and alpha radioactivity, 218n13; and Appendix C, 183–189; as clock for Earth's age, 37–38; discovery of, 1, 33, 43, 154; and discovery of radium by Marie and Pierre Curie, 33; and disintegration, 37; and Earth's age, 1, 33, 36, 37–38, 45, 46, xvi; exponential decrease of, 37; and periodicity of radioactive elements as constant, 44; and radiation from uranium salts, 33; and radioactive elements as a heat source, 34; and radio halos from alpha particles, 39
relativism, 155–157, 165, 168, vii, xiii–xiv
relativity, theory of, 3, 149, 150, 152, 183
relativity of motion, 141–143
religion, 82; and allegorical nature of Scripture, 122–123, 125; and calendars needed for festivals, 51, 75; and Christianity, 6, 52, 123, 125; and Church of England's attacks on Darwin, 29; and Counter-Reformation, 114, 122; and Creation, 3, 13–14, 127; and division between European science and Muslim science, ix; and division between reason and faith, 103–104, 237–238n55, x; and Earth's age, 1, 13; and end of Earth, 29; and evolution of church's relationship with science, 128–129; and Galileo on salvation, 121; and Inquisition, 115, 125, 238n57, xvi; and intelligent design, 127–128, 153, 156, 234–234n50, xiv; and interpretation of *The Bible*, 114; and Islam, 6, 52; and Jesuits, 9, 10; and justification of God in medieval period, 127; and monotheistic religions, 52, 126, 129–132; and mythology, 5; and obstacles to scientific knowledge, 49, vii; and Pope Benedict XVI, 128–129; and 'prime mover', 226n24; and Protestants, 122, 129, 130, 231n16; and relationship between science and faith, 131–133; and religious fundamentalism, 113, 156, vii, xiv; revival of, 201–202n1; and science, 157, 168; and Spinoza, 125, 126; and struggle against rationalism, 237–238n55; and texts by God vs. texts by Church Fathers, 122. *See also* Catholic Church; Christian world; creationism; intelligent design; Jewish world; Jews; Newton, Sir Isaac
Renaissance, 43; and astrology, 51–52; and astronomy, 104; and Bible as source of knowledge, 1, 7–8; and developments in firearms and printing, 7; and diagram of scientists of Renaissance to nineteenth century, xxii; and discovery of America, 7; and Earth's age, 1; and means of communication, 86; and scientific knowledge as part of culture, 7; and scientists abandoning literalist readings of sacred texts, 147; and scientists of as creationists, 156; and views on *The Bible* changing during, 43–44
retrogradation, 48–49, 55, 56–59, 61, 64, 69, 71; and explanation for planets' apparent motion, 82; of Mars, 221n8; and model of Nicolaus Copernicus, 79, 84; and periodic retrogradations, 72
Roentgen, Wilhelm, 33
Russell, Bertrand, 128, 156, 247n3, xiv. *See also Science and Religion* (Russell)
Rutherford, Ernest, 33, 34–35, xv; and discovery of atomic nucleus, 39; and method for dating of rocks, 37, 38

Salam, Professor Abdus, x
science: and abandonment of geocentrism

and development of heliocentric models, 48–49, 79–80; and absence of parallax, 105–106; and Arabic scientists rekindling Greek torch, 59; and astronomy, 154; and biology, 44; and birth of modern science, 59, xvi; and celestial geometry of Greeks, 48, 49; and challenge to scientific knowledge, 161–167, 169, xiii–xiv; and chaos theory, 227n41; and chemistry, 1, viii; and computer revolution, 99, 153; conservatism of, 150–151, 168, 248n7; and critiques of previous theories, 98, 148; and data, 229n2, viii; and dendrochronology, 45, 252n4; and determinism, 51, 100, 128, 149; and diagrams of eras of scientists, xix; and discovery of X-rays by Roentgen, 33; and Earth, 26, 147, 168, xiv–xv; and Earth sciences, xix; and electricity as changing understanding of world, 151; and faith, 131–133; and Foucault's pendulum in 1850, 144, 145; and general culture, 147, xvi–xvii; and harmony with religion, 128; and help of historians of science, 103, 155; and human scale, 152–153; and imagination, 168–169; and importance of research, 155–156; and Miletus school as origin of scientific thought, 63; and natural sciences, xv; and Nature, 161–162, 164; and Newton's all-encompassing theory of motion, 97; and Newton's contributions, 94, 96; and nuclear energy, 33–35; and nuclear physics, 3, x; and obscurantism, 103, 136, 155, 156; and observation and experimentation, 46, 136, 147, 148, viii; and palaeontology, 1, 27, 45, 154; philosophy of, 156, 161, 165, 166, x–xi; and physics, 1, 80, 98, 154, x, xix, xv–xvi; and quantum mechanics, 3, 128, 149, 150, 152; and Reich's experiment of 1831, 143, 144, 175–176; rejection of, 201n1, xiv; and relationship with the church, 128–129; and relative movement as opposed to absolute movement, 47–48; and relativism, 155–156, 157, 165, 168, 250n30; and release of heat associated with the disintegration of radium, 34; and religious fundamentalism, 113, 156, 157; and Richter's experiment with pendulum, 140; and scientific procedure, 21, 45, 147; and scientific truths, 2, 45, 147, 149–150, 166–167, xiii, xv, xvii; and scientists as authors, 103, 165; and search for truth compared to search for judicial truth, 149–150, 247n6; as separated from religion and politics, viii; social construction of, 155, 250n30; and social progress, 149, xiii; steps in development of, 104; and technoscience, 155; and tension between faith and facts, 159–160, 162–163; and theory of general relativity, 98; and thermodynamics, 3; and transition from myth to knowledge, 167–168; as universal, viii–ix; and unrecognized scientists Galois and Mendel, 150; and value of mathematics, 99–100; and view of Saadia Gaon on science vs. sacred text, 124; as Western imperialist construct, viii. *See also* Greek antiquity; Pakistan

Science and Religion (Russell), 128

seasons, 55, 56, 56 fig. 13, 66; and distance from Sun, 221n6; of equal length, 72; of irregular duration, 222n27

Smith, William, 17

Soddy, Frédérick, 33, 37, 39

Sokal, Alan, 155, 161, 162, 164, 166, 220n29

solar parallax, 229–230n3. *See also* parallax of stars

solar system: age of, 40; Babylonian models of, 60–61; and deterministic chaos, 102, 153; formation of, 2, 41, 109, 111, 153; geocentric models of, 77–78, 80; heliocentric representation of, 56 fig. 13; of planet Earth, 42, 46, 100, 119; and planets from other systems, 220n30. *See also* heliocentrism

solstices, 56, 63

spheres, 88 fig. 28; and circular motion of planets, 86, 93–94; cooling times for, 15, 35; decentring of, 225n8; displacement of planets on, 146; and end of 'corporeal spheres', 78; of fixed stars, 83; materiality of, 87; and model of homocentric spheres, 63–64, 65–67, 85; and Plato's model of the universe, 63–64; size of in Copernicus' model, 82

Spinoza, Benedict de, 125, 126, ix

stars: and aberration of light, 137–138, 144, 146; annual motion of, 53–55, 59; chemical composition of and spectroscopy, 154; composition of and centrifugal force, 83; and constellations, 53, 56, 57 fig. 14, 60, 64; and Creation, 3; distances of, 49, 105–106; diurnal motion of, 53, 59; and Doppler-Fizeau effect, 139–140; embedded and revealed phases of, 41–42; and fixed stars, 53, 55,

59, 64, 68, 71, 72, 81, 82, 83, 93, 137, 146; formation of, 41–42; and measurement of the angular distance between two stars, 72; and Milky Way, 95; movement of, 47, 56–59; number of, 152; parallax of, 81, 137, 138–139, 144, 146, 173; and Polaris, 24; and proto-planetary disc, 42; and supernovas, 41; and systems of planets, 41; and thermonuclear reactions of hydrogen combustion, 41; and Thuban, 24; and trajectory of the Sun known as the ecliptic, 55. *See also* parallax of stars; Sun, the

stratigraphy, 15–16, 17

Sun, the, 56 fig. 13, 166, xiii, xvi; age of, 20, 36; and ancient astronomy, 56, 57 fig. 14, 60, 63, 68, 69; angular speed of, 72; in *The Bible*, 114–115, 239n66; as cause of planetary motion, 90–91; as centre of Earth's annual orbital motion, 86; chemical composition of and spectroscopy, 249n21; as contemporary of Earth, 36; and cycles of thermonuclear reactions as energy source for, 39; deferent of, 72, 78; diameter of, 107–108; and Earth-Sun distance, 49, 73, 81, 105, 108–109; energy of, 20–21, 34, 36, 93; and Galilean frame of reference, 142; and gravitational collapse, 36; and heliocentric theory in *De revolutionibus orbium coelestium*, 81, 86; helium in atmosphere of, 218n2, 218n12; and hypothesis of Sun transferring magnetic energy which heated Earth, 28; and illumination of Moon, 53, 54 fig. 11, 55; influence of, 22; and length of days and years, 53, 59, xvi; mechanism of combustion of, 20; motive spirit of, 92–94; movement of, 47–48, 53, 55, 57 fig. 14, 59, 63, 70–73, 73 fig. 23, xvi; and nuclear fusion, 36; and 'nuclear winter', 26, 216n50; origins of, 3, 42, 44; and perihelion, 56, 72, 91; and period of rotation, 93; radius of, 108; as red giant, 30; and solar energy, 183–184; and solar parallax, 229–230n3; and solstices, 56, 63; spots on, 61, 95; and term "average Sun", 81, 84; and trajectory known as the ecliptic, 55, 56, 59 fig. 17, 63. *See also* heliocentrism; solar system

Swann's Way (Proust), 159–160

Syntaxis mathematica (Ptolemy), 73

Thomas Aquinas, St, 117, 122, 226n24, 237n54

Thomson, William. *See* Kelvin, Sir William Thomson

tides, 19, 22–23, 93, 118–119, 134, 136, 213n39, 232n17; and evidence of Earth's movement, 232n20; friction of, 214n42; increase in strength of, 214n43

time: and creation, 209n5; and Darwin's 'vast amount of time', 27, 33; definition of, 3, 30; determination of by atomic frequencies, 248n14; and geological time, 27, 35; and regularity of Earth's motion, 100; and sidereal year, 53, 61, 215n44; and solar year, 61, 100; and timescale from the Big Bang to the present, 2; and time taken for Earth's formation, 42–43. *See also* creationism; uniformitarianism

uniformitarianism, 24; and belief in long timeframes, 25, 27; geologists' acceptance of, 28; and permanence of the laws of physics, 26

United States: and creationism, 29, vii; and religious sects, 202n6

universe, 7; and absolute space, 145, 146, 219n16; and Big Bang theory, 2, 3, 40, 45; and centre of in Copernicus' theory, 81; chronology of, 2, 2 fig. 4, 3, 40; and dark matter and energy, 152; date of, 3, 40; and deterministic chaos, 101–102, 153; and divine intelligence, 101, 102; and Hubble's rate of expansion, 40; and impossibility of predictions, 101–102; as infinite and without centre, 115, 116; measurements of distances and dimensions of, 49; and Milky Way, 95; and model of Miletus school, 63; and movement of matter on a large scale, 40; and philosophy, 114, xi; red shift of galaxies indicating expansion of, 139

Ussher, Archbishop James, 43; and the *Bible*, 44–45; as biblical chronologist, 205n19; and Earth's age, 8

Verdet, Jean-Pierre, 103, 223n37, 228n47